Catholic Schoolhouse

Year 2 Science

Botany, Chemistry, Astronomy, Sound & Light

by Kathy Rabideau

Year 2 Science
©2015 Kathy Rabideau

www.catholicschoolhouse.com

Edited by Jeanne Konwal

Printed in the United States of America

Although the author has made every effort to ensure that the information in this book was correct at time of press, the author does not assume and hereby disclaims any liability to any party for any loss, damage, or disruption caused by errors or omissions, whether such errors or omissions result from negligence, accident, or any other cause.

Permission is given to the purchaser of this book to make copies of the printouts for use in your classroom or group. No other part of this book may be copied or reproduced by any means.

ISBN 978-1-4675-3407-9

Cover design by Kathy Craig Clark

To the Teacher

Catholic Schoolhouse *Year 2 Science* offers hands-on exploration of Botany, Astronomy, Chemistry, and Sound & Light. Catholic Schoolhouse Year 1 and Year 3 complete a comprehensive three-year cycle of science topics, ensuring a well-rounded education.

Having a few established rules for science class will not only help class go more smoothly, but will also be a valuable way for the students to learn self-control.

- No eating or drinking in a science lab—ever! We are setting up a real science environment as much as possible.
- When you arrive at class, no touching until given permission by the tutor. (This is more difficult and more valuable than you might imagine!)
- Raise your hand to speak.
- Stay seated unless instructed to move around the room.
- Attention should always be on the tutor, or you might miss something valuable.
- Most important, have a good attitude. Science has a lot of exposure to slimy, creepy, new things. When you say "ewww," or "gross," or other negative sounds, it brings down the whole class. To establish a positive learning environment, the tutor must not allow this. A great trick we learned from a former teacher is to teach that "ewww" (scrunch nose if necessary) is the sound of a closed mind. "AAhhhh" is the sound of an open mind, eager to learn. Have the whole class practice saying "ahhh" as often as you hear negativity. It works great!

Special terminology is bolded throughout this lesson plan. Extra attention should be given that students become familiar with these terms. The tutor should repeat these terms frequently while explaining the lesson. Asking students to repeat these words is also beneficial. Quizzing after class can create productive competition and opportunities to praise students for good attention and a job well done in class.

Often, science has assembly work that can be done ahead of time for the youngest students. For this reason, you might wish to schedule your oldest students to come to science first. Then, all the prep work is done, and class is ready for the younger students who might work a little more slowly. Teaching the older students first also makes it easier for the tutor to gauge responses to bring the lesson down to the younger students' level appropriately.

You might wish to have students keep a "lab manual" to record what they learn and observe during the year. Writing or drawing what they have learned cements the lesson. Research can be done and extra information added to the lab book to broaden the student's knowledge. Definitions of the bolded terminology could also be included.

Remind your students daily that they are great scientists, and have a fun year!

Science Supply List

Quarter 1 - Botany

Week 1
- **Celery**—one stalk per student in class; stalks can be cut to half height.
- **White carnations or daisies**
- Scientific method **Lab Report sheet**—one per student (in Appendix)
- **1 clear glass** or jar of water to demonstrate lab setup
- Red and blue **liquid food color**
- **Paper towels**
- **Pencils**
- **Colored pencils**

Week 2
- **Organic spinach leaves**
- **Distilled water**—100 ml or about ½ cup
- **Baking soda**—¼ **teaspoon**
- **¼ teaspoon measuring spoon**
- **Syringe with no needle**
- **Hole punch**
- **Lamp with incandescent lightbulb**
- **Beaker, large graduated cylinder, or clear glass jar**
- **Metal butter knife**

Week 3
- A variety of real or artificial **flowers**—one per student. These must have a very clear stamen and pistil for students to be successful.
- **Magnifying glasses**—one per student in largest class
- **Tweezers**—optional, to feel like scientists!
- Additional specimens of a different type for older students
- **Paper**—½ sheet per student
- **Pencils**, colored pencils, or markers for labeling if time allows.

Week 4
- **Samples of seeds** using the various methods of dispersal
- Man-made counterparts (optional): **Velcro, plastic helicopter**
- **Plastic standard CD cases**—one per student, used or new, but must be clear, not slim.
- Bag of **potting soil**
- Several **bowls** in which to put potting soil—one per two students in class
- One **spoon** per student in class
- **Bean seeds**—three per student
- Large **paper plates** to catch dirt

Week 5
- Variety of **leaves,** including **complete branches,** with each of the above characteristics
- **Scavenger hunt printouts**—in Appendix

Week 6
- Tree **dichotomous key clue sheets** can be downloaded from www.catholicschoolhouse.com
- **Tape** to hang dichotomous clues around room.
- **Assorted leaf samples** from clue sheets for identification—at least one per student

Quarter 2 - Chemistry

Week 7
- **Graduated cylinders**—50 ml is a good size, but a variety is fine. One per student or pair of students in class.
- **Medicine droppers**—one per student, or pair of students in class.
- **Water**
- **Liquid food coloring**
- **Paper cups**
- **Paper towels**
- Larger **cups** or **pitchers** to pour waste water in.
- **Pencils**
- **Paper**—¼ sheet is fine.

Week 8
- **Atom cards**—print from Appendix, one sheet for every two to three students in class
- **Molecule cards**—print from Appendix, one set per student in largest class

Week 9
- **Candle** and **match** or **lighter**
- **Chemically activated ice pack**—sold in first-aid section
- **Borax**—½ teaspoon per student
- **White Glue**—⅛ cup (30 ml) per student
- **Water**—80 ml per student (⅓ cup per student)
- **Food coloring** (optional)
- **Zip-top bags**—sandwich bags work fine, one per student
- **Graduated cylinders**
- ½ teaspoon measuring **spoons**—two per class is fine
- **Stir sticks**—popsicle, coffee straws, or plastic spoons
- **Paper cups**—one per student

Week 10
- **Purple cabbage juice** (made from red or purple cabbage)
- **Lemon juice**—20ml per student
- **Cheap shampoo (the most alkaline)**—20ml per student
- **Baking soda**—1-pound box per 30 students (3 teaspoons per student)
- **Medicine (eye) droppers**—one per student
- **Graduated cylinders**—one per student
- **Florence flasks** or pint Mason jars for mixing—two per student (plastic cups will work in a pinch)
- **Plastic teaspoons**—two per student
- Litmus paper (optional)—two per student
- **Paper towels**
- **Small paper cups** - three per student (lemon juice, shampoo, baking soda)

Week 11
- **Mixture of house items in large bowl** to demonstrate heterogeneous mixtures (rice, beans, coins, Legos, cotton balls, and so on)
- **Coffee filters**—one per student
- **Glass pint jars**—one per student in largest class
- **Black Crayola marker or similar washable markers**—*Sharpies do not work*. Two to three per class.
- **Scissors**—two to three pair per class should be plenty.
- **Water**—one cup per student
- **Paper towels**—one per student

Week 12
- **Tall, clear container - carafe, vase or quart mason jar**—one per class demo
- **Heavy solids**—washers, pebbles, marbles...
- **Small piece of wood**—make sure it floats
- **Water**
- **Ice cubes**
- **1 pint mason jar**—for corn oil/ice cube demonstration
- **Graduated cylinders**—one per student in class
- **Corn oil**—⅛ cup per student
- **Corn syrup**—⅛ cup per student
- **Water**—⅛ cup per student
- **Paper cups**—three per student to distribute oil, corn syrup, and water

Quarter 3 - Astronomy

Week 13
- Sun—a **balloon** that can be blown up to an 8-inch diameter **or yellow ball**
- Mercury—a candy **sprinkle** used to decorate cakes, 0.03-inch diameter
- Venus—a **peppercorn**, 0.08-inch diameter
- Earth—a second **peppercorn**
- Mars—a second **sprinkle**
- Jupiter—a **pecan, walnut, or chestnut**, 0.9-inch diameter
- Saturn—a **filbert or acorn**, 0.7-inch diameter
- Uranus—a **peanut or coffee bean**, 0.3-inch diameter
- Neptune—a second **peanut or coffee bean**
- Pluto—a third **sprinkle**

Week 14
- **Colored paper** (and labels for the youngest children) as follows:
  Core, 27,000,000 °F, **white** (3" circle)
  Radiative zone, unknown, **pale yellow** (4.5" circle)
  Convective zone, varies, **yellow** (6" circle)
  Chromosphere, 6,000-500,000 °F, pale **yellow** (8.5" circle)
  Photosphere, 10,000 °F, **dark yellow/gold** (7.5" circle)
  Sunspots, 4,000 °F, **brown** (random blobs)
  Solar flares, **orange** (hollow loops, cut 2" long to have 1" gluing surface, 1" extension)
  Corona, 2,000,000 °F, **white** (white branching, feathery shapes, up to 5" long)
- **Brass fasteners**—one per student
- **Scissors**—one per student in class
- **Glue sticks** and **preprinted labels with above zones** OR
  Pencils for writing zones on the sun

Week 15
- **Globe** that can be carried by students
- **Lamp** with shade removed

Week 16
- **Space Grid**—in Appendix, one per student
- **Coordinates of various constellations**—in Appendix, print and/or write on board.
- **Pencils**—one per student with erasers
- **Heavy black paper or cardstock**—one to two per student
- **Pins** to poke holes—one per student in class
- **Scotch tape**

Week 17
- **Super 6 rocket kit**—one per student, plus a couple of extras from http://www.pratthobbies.com/
- **Rocket engines**—one per younger student; two per older student
- **Launch pad and ignition system**
- **White glue**
- **Paper towels**
- **Markers and highlighters** for decorating rockets

Week 18
- **Rocket launching equipment—pad and igniter**—see week 17.
- **Rocket engines**—one to two per student
- Spare **batteries** if your igniter is battery-powered
- Students' completed **rockets**

Quarter 4 - Sound & Light

Week 19
- **Polarizing light sheets**— two 2"—3" squares per student, cut from large sheets
- **Light source**—flashlight or lamp
- **2 ropes**—jump ropes work fine.
- **2 poles**—broom handles work fine.

Week 20
- **Lenses**—an assortment, or one concave and one convex lens per student in class
- **Mirrors**—one per student in class
- **2 Flashlights**—LED works best because it has a more focused beam.
- 2 pieces of white **cardstock** for mirror game targets

Week 21
- **Prisms**—several, one per student in class, if possible.
- **Light source**—direct sun from a window will work, but going outside would be better.
- **White paper**

Week 22
- **Rope**—a jump rope will work fine.
- **Coil spring**—Slinky is perfect.
- **Wide glass bowl**
- **Plastic wrap**—good quality
- **Rice**—about 1 tablespoon

Week 23
- **Flat rubber bands**—⅛" and ¼" wide, one of each per student in class
- Sturdy **plastic cups**—one per student in class
- **Straws**—one per student
- **Wineglasses**—8-10 to get a good variety of pitches
- **Water**

Week 24
- **Plastic drinking cup**—one per student
- **Yarn or cotton string** (nylon string will not work well)—one 20" per student
- **Nail**—two to three per class
- **Scissors**—one per student
- **Paperclip**—one per student
- **Water** in a small bowl—one per class
- **Paper towels**—one per student

Plant's Subway System
Week 1

Plants pick up nutrients and water through their roots. They move up the stem through xylem and phloem, then transpiration releases water through the leaf's stomata.

How do plants transport nutrients from their roots to their leaves?
What chemical makes plants green?
What is photosynthesis?

Introduction

Plants produce food through **photosynthesis**. Large amounts of water, with its dissolved minerals, are needed for this chemical process. For this to happen, plants must have a way to get the water all the way up to their leaves.

Food and **water** are transported through the plant through structures called **xylem** and **phloem**. A plant doesn't have a heart to pump liquids as we do. Our heart *pushes* the blood through our body. In plants, water is *pulled* up from the roots to the leaves. The two processes that make this happen are **transpiration** and **cohesion**.

Transpiration: water escapes through tiny openings on the bottom of the leaves, called **stomata**. This creates a difference in pressure in the **xylem** cells. When one molecule is lost, another is pulled up to replace it. A similar thing happens when you remove water from the top of a straw—other water molecules must come up the straw to replace them. Leaves constantly release water through **transpiration**.

Cohesion: cohesion is the force that cause water molecules to stick to one another—the reason drops gather together. Because xylem vessels are very small, cohesion causes water molecules to be pulled along in a process called **capillary action**.

Xylem vessels connect straight from the roots to the leaves for the speediest transport possible. The rings you see when you cut down a tree are old xylem tissue. In fact, the word *xylem* comes from the Greek word *xylem*, meaning "wood."

Supplies

- **Celery**—one stalk per student in class; stalks can be cut to half height.
- **White carnations or daisies**
- Scientific method **Lab Report sheet**—one per student (in Appendix)
- **1 clear glass** or jar of water to demonstrate lab setup
- Red and blue **liquid food color**
- **Paper towels**
- **Pencils**
- **Colored pencils**

Teacher Prep

The night before class:
- Place celery in colored water. Having some leaves attached is dramatic.
- Split the stem of a carnation or daisy. Place one end in red water, place the other end in blue water.
- Keep one celery stalk and one flower uncolored for class.

Procedure

- Beginning with guidelines will help you to set the tone for a stress-free environment.
  Go over the rules in the front of this book on the To the Teacher page and share them with the students. Using humor and animation will make class more fun—giggles are always OK! Especially when you discuss no eating in science lab—ask them if it will be OK to eat the celery—NO! Drink the water? No, that is so silly!
- In science class today, we will learn about the scientific method while studying how nutrients flow through plants.
- Explain the steps of the scientific method: question, hypothesis, experiment, observation, analysis, and conclusion.
- Show an uncolored stalk of celery and a carnation to the class. "Can you see the xylem in these plants?" (probably not)
- "Let's devise an experiment that will allow us to observe the xylem."
- After brainstorming for a while, lead students to the idea that because xylem transport water, if we put dye in water, it might make the xylem visible. Perhaps a hint about how medical dye is used to make our blood vessels visible to medical scans such as an MRI.

- Record the question, "How does water travel through plants?" Students may write or draw according to their abilities.
- After the procedure is decided, bring out cup of water, add food coloring, place celery in cup.
- Have students explain or draw the lab set-up under procedure.
- "We would normally have to wait for the water to be transported. I put one in water last night so we could observe our results right away like in a cooking show!"

- Bring out changed celery in colored water. Remove stalks from water, pat dry on paper towel and pass to students for observation.
- Make sure students observe the dots on the bottom of stalks—these are the ends of the xylem.
- Observe the leaves also. Do you see any xylem in the middle? Why? (perhaps it is beneath the outer "skin.")
- "Is the xylem branching, like our circulatory system, or does it go straight up the stalk?" (straight, no branching)
- "Can anyone think of a way we could test this?" (Split the stem, and see where the color goes.)
- After discussion, bring out the carnation with the split stalk. The carnation should be half red and half blue.
- "What is your conclusion?" (The color goes straight up, no horizontal mixing. Xylem vessels are straight, not branching like our circulatory system. If you cut into the trunk of a tree beyond its xylem, the part of the tree above the injury will die.) Complete lab sheets by filling in conclusion.

"We cannot create observers by saying 'observe', but by giving them the power and the means for this observation and these means are procured through education of the senses."

Maria Montessori

Photosynthesis
Week 2

*Plants need chlorophyll to make their own food with photosynthesis.
Through respiration, plants use the food; it's really as easy as this.
They use carbon dioxide to give off oxygen, making the air we breathe.
These processes that plants must do happen in the leaves.*

What is necessary for photosynthesis to occur?
What is released during photosynthesis?
What color is chlorophyll?

Introduction

Photosynthesis is the process by which a plant uses light, water, carbon dioxide, and **chlorophyll** to make its food. Oxygen, which the plant does not need, is released during the process. This is the chief purpose of leaves.

All food we eat starts with photosynthesis.

Plants are green because of the chlorophyll the leaves have. The parts of the leaf where the chlorophyll is stored are called **chloroplasts**. You cannot see photosynthesis as it occurs, but in our experiment today, we will see the effects of photosynthesis. The oxygen released during photosynthesis gathers on the leaf surface, making the leaf pieces float.

Supplies

- **Organic spinach leaves**
- **Distilled water—100 ml or about ½ cup**
- **Baking soda—¼ teaspoon**
- **¼ teaspoon measuring spoon**
- **Syringe with no needle**
- **Hole punch**
- **Lamp with incandescent lightbulb**
- **Beaker, large graduated cylinder, or clear glass jar**
- **Metal butter knife**

Teacher Prep

- Today's experiment is a demonstration.
- Set up your science lab so all students can see what is happening.

Bonus

Plant respiration can be observed by placing a leaf inside a plastic bag in a sunny window. As the plant gives off water, you will see the water collect as droplets inside the bag. Compare rates of respiration by placing bags in different locations.

Week 2 - Botany

Procedure

- "Today, we will watch photosynthesis happen. Do you think this is possible?" (no, but we will see its effects.)
- Fill beaker or graduated cylinder to the 100 ml line with distilled water. (If using glass jar, pour in ½ cup water.)
- "What do plants use during photosynthesis?" (carbon dioxide and light)
- We will use baking soda for our carbon dioxide source.
- Add ¼ teaspoon baking soda and stir.
- Create circles of leaf with a hole punch, avoiding the veins if possible.
- Students can be allowed to each punch one circle if desired.
- Put the circles in the syringe.

- Draw up enough water to fill the syringe ¾ full.
- Squeeze out the air.
- Now, we need to remove the air from the leaves.
- Seal the tip with your finger and pull the stopper to create a vacuum while a student helper taps the side of the syringe with the knife. This forces the air out of the leaves and helps the bubbles float to the top.
- Remove your finger and squirt out the bubbles and air at the top of the water. (Some water will squirt out too.)
- Repeat this process until the leaf pieces sink.

- Remove the stopper, and pour the leaves and water into the beaker.
- We now have leaves with their oxygen removed in a water solution containing carbon dioxide.
- "What else do we need for photosynthesis?" (light)

- Set the container with the water/baking soda mixture and the sunken leaves under a lamp.
- Explain photosynthesis to the students while waiting for the first leaf to lift.

- After a few minutes (3-8), you will notice the first leaf lift one side off the bottom.
- Watch closely, as once it begins, it floats quickly to the top.
- Continue watching as more leaves float to the top, demonstrating that photosynthesis is occurring and oxygen has been made.

Did you Know?

In the **fall**, the shorter days signal trees to produce less chlorophyll. As the **chlorophyll** fades from the leaves, the **yellows**, **oranges**, and **reds** become visible. Small amounts of these colors were there all the time; they were just covered by the strong green pigment in the chlorophyll! **Cooler temperatures** cause other changes that add more **vibrant** colors to the leaves.

Dissecting Flowers
Week 3

Pollen from the stamen is moved to the stigma.
This is pollination.
A tiny seed forms, so new plants can grow,
and the cycle goes on and on.

What are the four main parts of a flower?
When does pollination occur?
Which two parts of the flower are actually modified leaves?

Introduction

All plants produce **flowers** for the same reason—to produce seeds. **Seeds** are necessary for the plant to reproduce. The most important parts of a flower are the **stamen** and the **pistil**. The stamen, composed of the **anther** (top) and the **filament** (tube), produces the **pollen**. The **pistil** is the flower's female reproductive organ containing the ovary. It is composed of the **stigma** (top), the **style** (tube), and the **ovary** (base). If you research, you will find a great variety in the naming of flower reproductive parts. For example, sometimes the pistil is called the stigma, other times, it is presented as one part of the stigma. We have presented those most common in elementary texts.

For a seed to be created, the **ovary** must be **fertilized**. **Pollination** occurs when the pollen is transferred from the stamen to the pistil and makes its way to the ovary. After pollination, the seed begins to develop. A fruit often forms from the ovary around the fertilized seed.

In its most basic shape, a flower is made of four concentric parts. There is an outer ring of modified leaves called **sepals**. These protect the flower before it opens, and they are usually green. Inside the sepals is another ring of modified leaves called **petals** that are often brightly colored. The **stamens,** the male reproductive structures, contain pollen. In the very center of the flower is the **pistil**. This might look like one tube, many tubes like a sunflower head, or a swollen ovary, indicating that the flower is already fertilized.

Supplies

- A variety of real or artificial **flowers**—one per student. These must have a very clear stamen and pistil for students to be successful.
- **Magnifying glasses**—one per student in largest class
- **Tweezers**—optional, to feel like scientists!
- Additional specimens of a different type for older students
- **Paper**—½ sheet per student
- **Pencils**, colored pencils, or markers for labeling if time allows.

Procedure

- Explain that **botanists** are scientists who study plants. Today, we will become botanists by studying a part of a plant—the flower.
- Review the rules of science as listed in the introduction. This will be their first introduction to the self-control they need in science class. Ask them to practice self-control by *not* touching the magnifying glasses.
- Place magnifying glasses on table in front of each student.
- Pass out flowers. Again, they need to practice self-control. If someone touches, don't be harsh, just gently say, "Oops, we need to go over our science rules; can you all tell me what they are?" ("No touching" should come up.)
- When all are seated, without touching, waiting patiently, tell them it is time to begin. They may pick up their magnifying glass and touch their flower.
- Encourage students to examine their flower without damaging it. Help them identify the main parts—sepal, petal, stamens, pistil. Older students may be capable of more detail.
- Pass out tweezers. Carefully remove parts of your flower.
- Help students correctly identify common parts.
- If time allows, glue parts on paper, and label them. You may also give older students an additional different specimen to examine. Does it have the same parts? How are they alike? How are they different?

Easter Lily

Yellow Ray Flower

Sepal (1) *Green part just below the petals*
Petal (2) *Colorful part of the flower*
Stamen (3) *Fuzzy texture of pollen helps to locate these, usually on the outer ring*
Pistil (4) *Center tube(s) that receives the pollen. If swollen, it is already fertilized and is the ovary.*

Bonus

Keeping a nature journal is a great way to record the flowers you see. Use your best art skills to draw the flowers, then don't forget to write about the flower and how you feel about it! If you start this week, you can journal your way through the entire botany study. *Handbook of Nature Study* by Anna Comstock is a well-loved resource available for purchase or free to download on the web.

"Let them once get in touch with nature and a habit is formed which will be a source of delight and habit through life."
—Charlotte Mason, *Original Homeschool Series.*

Traveling Seeds
Week 4

Four types of seed dispersal: water, animal, wind, mechanical

What are some ways that plants disperse their seeds?
What inventions mimic seed dispersal techniques?
Have you ever helped transport a seed?

Introduction

Plants have limited mobility, so they must use creative methods to spread their seeds. This is important if the plant is to grow in new areas. **Plants use animals, wind, water, and even more surprising methods to disperse their seeds**.

Animals transport seeds when they eat the fruit, then leave the seeds as waste in another place. Birds disperse many seeds over long distances in this way. Squirrels bury their nuts giving the new seedling a head start. Animals also carry seeds that attach to their fur.

Wind can blow seeds that have sails, such as dandelions. The shape of seeds can also help seeds such as the maple tree "helicopters." move from the parent tree.

Water—Coconuts can float on water for many miles until they land on a fertile beach. Some water plants disperse their seeds directly into the water to float to their new home.

Mechanical dispersal occurs when plants forcibly eject their seeds. The seeds sometimes shoot out of their pods with great force. The names of the squirting cucumber and snapweed give a clue about their seed dispersal method.

Some **inventions** of man were first inspired by seed transportation. Velcro uses the same technique as the hooks in burdock. Windmill blades are shaped much like maple tree "helicopters."

Supplies

- **Samples of seeds** using the various methods of dispersal
- Man-made counterparts (optional): **Velcro, plastic helicopter**
- **Plastic standard CD cases**—one per student, used or new, but must be clear, not slim.
- Bag of **potting soil**
- Several **bowls** in which to put potting soil—one per two students in class
- One **spoon** per student in class
- **Bean seeds**—three per student
- Large **paper plates** to catch dirt

Procedure—Part 1

- Have seeds on table when students arrive. As they enter, ask them to share the science class rules.
- Discuss the various ways seeds are dispersed.
- Show any man-made counterparts you might have.
- Ask students to brainstorm inventions that might use ideas similar to the way seeds travel.

Procedure—Part 2

- After seeds are dispersed, they must find fertile soil in which to grow. We will observe a seed sprouting and growing by making a take-home plant viewer.
- Pass out paper plates, bowls, and spoons. "Remember, self-control . . ."
- Scoop about ½ cup of potting soil into each bowl. If you wish to add humor to your class, you can emphasize "no eating." This should result in some giggles!
- Pass out CD cases and beans—instruct students to carefully place a few spoons of potting soil on lid of case and place three beans in potting soil. Carefully close the CD case.
- At home, students should keep the soil moist, but not wet.
- Have students watch their beans sprout to see which way the sprout goes. **Geotropism** causes the roots to go down and the leaves to grow up. You could have some students place theirs upside down to see what happens!

Bonus

Try to collect as many different seeds as you can. Add these to your nature journal. Be sure to draw the plant you think they came from and describe what it was like to find them.

"There is no part of a child's education more important than that he should lay — by his own observation — a wide basis of facts towards scientific knowledge in the future."

—Charlotte Mason

Leaf Scavenger Hunt
Week 5

Two kinds of leaves, compound and simple, are arranged alternate, opposite, or whorled.

What are three examples of leaf arrangement?
What are leaf veins that fan out like a palm called?
Leaves that contain more than one leaflet are known as _____?

Introduction

Characteristics of leaves are used by scientists to classify and bring order to the understanding of the plant world.

Simple leaves are individuals, occurring alone, while **compound leaves** have leaflets or parts that make up the leaf.

The leaves, whether simple or compound, can be arranged on the stem in different ways:
Alternate—arranged alternately or staggered on the branch
Opposite—leaves are attached directly across from one another
Whorled—arranged in a ring shape from a center point

How the veins are arranged is another category:
Parallel—veins running the same direction, like railroad tracks. These do not need to be mathematically parallel to be considered, only suggesting parallel.
Pinnate—a center vein, with secondary veins attaching opposite each other
Palmate—branching outward, like the palm of a hand

Scientists also describe the margins, or edges of leaves, as smooth, lobed, toothed, or serrated.

These are the basic classifications. There are many more variations as well as subcategories of the above that can be learned!

Supplies

- Variety of **leaves**, including **complete branches**, with each of the above characteristics
- **Scavenger hunt printouts**—in Appendix

Teacher Prep

- Select a sample of each type of leaf to show students.
- Spread remaining leaves around classroom for students to find.
- Alternately, head outside to gather leaves of each type if weather permits.

Week 5 - Botany

Procedure—Part 1

- Introduce the idea of types of leaves by showing several examples, then ask students to generate a definition of each. Discovery of scientific definitions is great fun.
- "Let's begin by learning one classification of leaves. This pertains to how they are arranged on the branch." Show samples of alternate, opposite, and whorled. Make sure they understand the difference between multiple leaves on the branch and compound leaves.
- For older students, explain the three types of veins: parallel, pinnate, palmate. Show a leaf with each type of vein. See if they can correctly match the name to the leaf. Give hints as needed. Palmate: like a palm. Parallel: like a railroad track. Pinnate: by process of elimination.
- "Now, let's play a game. I'll hold up a leaf, and you tell me as many labels as you know."
- Hold up samples according to their age level. Have fun trying to trick the students—compliment them when they aren't fooled!

Procedure—Part 2

- Now, we will see how many examples you can find. Hand out scavenger hunt sheets. Give students boundaries where they may search, whether indoors or outdoors.
- Spend 5—10 minutes searching.
- Come back together for the last 5 minutes to give students a chance to share what they found.

ALTERNATE
leaflets arranged alternately

OPPOSITE
leaflets in adjacent pairs

WHORLED
rings of leaflets

PINNATE
secondary veins paired oppositely

PARALLEL
veins running the length of the leaf, not intersecting

PALMATE
veins diverging from a point in a branching pattern

Bonus

Making a leaf poster is a great way to learn about leaves. See how many different examples you can find and label them accordingly.

Did you Know?

The **compound leaves** of **poison ivy** consist of a **center leaflet** flanked by two **oppositely arranged leaflets**. The edges can be **smooth** or **mildly toothed, not serrated**. The leaves often have a distinctive **"thumb"** protruding from the outer edges. The distinctive **red stem**, or **red color in fall**, can help confirm your identification.

Mystery Trees
Week 6

There are two main types of trees:
Evergreen trees, like pine and fir,
which stay green all year long
and deciduous trees, like maple and apple,
which lose their leaves in the fall.

What is the difference between deciduous and evergreen?
Can you name 3 evergreen trees?
How does a dichotomous key work?

Introduction

In botany, an **evergreen** plant is a plant that has leaves in all seasons. This contrasts with **deciduous** plants that completely lose their foliage during part of the year, usually the winter or dry season.

Common **evergreens** include most **conifers**, **live oaks** that grow in the southern United States, and **hollies** that may be deciduous in the far north. **Evergreen** trees lose their leaves, but not all at the same time. They replace them gradually over the year, and thus look green all the time.

Deciduous trees respond to changes in temperature, rain, and length of day to signal them to shed their leaves. Transpiration slows, preserving water in the tree, making the tree better able to survive harsh conditions. Dropping their leaves also helps prevent damage to their xylem from harsh weather.

Interestingly, things other than trees can be *deciduous*, a word that means to shed. Thus, deer have deciduous antlers, and you have lost your deciduous (baby) teeth!

A **dichotomous key** is a tool that allows the user to determine the identity of items. *Dichotomous* means "divided into two parts" from the Greek *dicho* meaning in two parts. *Tome* means to cut. Therefore, dichotomous keys always give two choices in each step.

Supplies

- Tree **dichotomous key clue sheets** can be downloaded from www.catholicschoolhouse.com
- **Tape** to hang dichotomous clues around room.
- **Assorted leaf samples** from clue sheets for identification—at least one per student

Teacher Prep

- Hang printed clue sheets around the room with plenty of space to facilitate movement by the students. If these trees are uncommon in your area, you might need to add pages. Prepare a chart on the board for students to record their identified trees. Chart should have three columns: Mystery leaf sample, Order of clue sheets used, Identity of mystery tree.

Procedure

- Welcome children, and explain that today we will work as botanists to classify leaf specimens.
- We will use a dichotomous key. Review how a dichotomous key works: you read the question, then follow the steps to the next question. By the process of elimination, you identify your specimen.
- Explain that today's dichotomous key is a series of steps around the room.
- Walk through sample classification, always beginning with sheet #1. Then, walk to the next question as directed on the sheet.
- Hand out labeled leaf specimens, one per student to begin.
- "Now, it's your turn; always begin with sheet #1. You might need an orderly plan to get them started, but then, they should be able to walk around the room as desired. Younger students might only get one leaf classified. Instruct older students to enter their results on the board and choose another specimen as time allows.
- Bring class together to discuss results. Was it easy? Was it difficult to tell? Did everyone who classified the same leaf get the same result? Would you like to be a botanist?

Trees

I think that I shall never see

A poem as lovely as a tree.

A tree whose hungry mouth is prest

Against the sweet earth's flowing breast;

A tree that looks at God all day,

And lifts her leafy arms to pray;

A tree that may in summer wear

A nest of robins in her hair;

Upon whose bosom snow has lain;

Who intimately lives with rain.

Poems are made by fools like me,

But only God can make a tree.

—Joyce Kilmer

Bonus

Making a leaf poster is a great way to learn about the trees growing near you. You can arrange them by evergreen and deciduous, then go further and label leaf venation, edges, simple and compound. Add a poem about a tree!

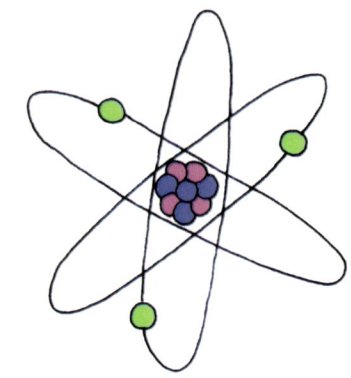

Graduated Cylinders and the Meniscus
Week 7

Chemistry is the study of matter and how it behaves. Atoms are the building blocks of matter, made of protons, neutrons, and electrons.

What is matter?
What are the three parts of an atom? Which are in the nucleus?
What is a meniscus?

Introduction

Matter, matter everywhere.
There's matter in your hair.
Matter in the air.
There's even matter in a pear!
There's liquid matter, solid matter, and matter that's a gas.
Even you are matter, because you have volume and mass! —*Steve Tomecek*

To scientists, everything in the universe is either **matter** or **energy**. **Matter** is all the "stuff" in the world, whether you can see it. **Energy** is what moves or changes the "stuff."

All matter is made of atoms, which combine to make **molecules**. Atoms contain **protons**, **neutrons**, and **electrons**. The neutrons and protons are in the nucleus and contain most of an atom's mass. The electrons orbit around the nucleus. You may have heard of even tinier parts called nucleons and quarks. Scientists are always working to find ever-smaller pieces of matter to understand our world.

All matter has volume and mass.
Volume—The amount of space taken up by matter. Even invisible gases take up space. When you blow into a balloon, the balloon must get bigger.
Mass—Mass is the total quantity of matter contained in an object.

Supplies

- **Graduated cylinders**—50 ml is a good size, but a variety is fine. One per student or pair of students in class.
- **Medicine droppers**—one per student, or pair of students in class.
- **Water**
- **Liquid food coloring**
- **Paper cups**
- **Paper towels**
- Larger **cups** or **pitchers** to pour waste water in.
- **Pencils**
- **Paper**—¼ sheet is fine.

Teacher Prep

- Make three different colors of water: red, blue, and yellow.
- Set out a half-full paper cup of each color and a graduated cylinder on a folded paper towel for each student or pair of students.
- Have an extra pitcher of each color available for refills.

Procedure

- Today we will learn an important skill used by scientists—that of measuring accurately and correctly. "Why do you think this is important to scientists?" (to repeat experiments, to tell others, to get precise results . . .)
- "Great job not touching!" "What do you see before you?"
- If no one has a ready answer, explain that this is a graduated cylinder. It has not graduated from college, but rather, in this case, *graduated* means "marked off in small measurements." Scientists use graduated cylinders for important measurements because the small surface area allows for precise measurement. Smaller cylinders are available for even smaller quantities.
- Hold up the graduated cylinder to show lines. Explain how to read the lines and have students locate a sample line, perhaps 20 ml.
- Have students pour a little yellow into their graduated cylinder. "Look closely at the surface of the water. This is called the meniscus." Have students repeat word.
- "What is it shaped like?" (a bowl or dish)
- "This creates confusion! If I want to record how much liquid I have, the measurement will be different if I look at the top or the bottom of the meniscus."

- Give students time to notice this. Then, explain that scientists have agreed that all measurements should be taken at the *bottom of the meniscus*.
- Ask students, one at a time, how much water is in their graduated cylinders.
- Instruct students to fill their cylinder with yellow to the 10 ml line. Use the medicine dropper to get the bottom of the meniscus exactly on the line.
- "Now, fill to the 20 ml line with blue. Again, be precise."
- Do not stir! Watch as the colors mix without stirring. This is called **diffusion** and happens because the individual atoms are always moving. "What color did you make?" (Green —yellow and blue make green.)
- "Now, I will allow you to make a color of your own; the only rule is that you must write the exact formula in milliliters (ml)."
- Pass out paper slips and pencils.
- After students make their colors and show their classmates, pick one to imitate.
- Have students dump out their cylinders and use the chosen formula. Do they all end up the same? They should if we are good scientists.
- If time allows, choose another formula to imitate or make new colors. Practice makes good scientists, plus they are learning about color theory.

Bonus

Dropping a drop of food color directly into water allows you to observe molecular motion. Make sure the water is still before beginning. Compare the motion of different temperatures of water.

"I consider nature a vast chemical laboratory in which all kinds of composition and decompositions are formed.

—Antoine Lavoisier
(1743-1794)
Catholic Scientist
"Father of Chemistry"

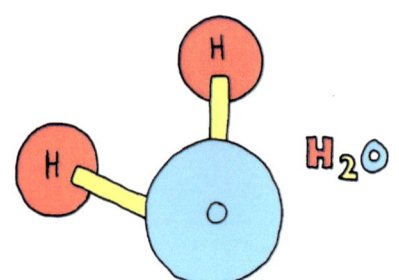

Let's build a molecule!
Week 8

*Each type of atom is an element and has a special symbol,
Like H for hydrogen, O for oxygen, and Au for gold.
When atoms combine, they form molecules,
like H_2O—that's water.*

Who first ordered elements in the periodic table?
What are molecules composed of?
What does the atomic number tell us?

Introduction

The **periodic table** we use today is based on the one Russian scientist Dmitri **Mendeleev** devised and published in 1869. With his periodic table, Mendeleev could predict the properties of undiscovered elements.

The periodic table is arranged in order by **atomic number**. The atomic number is the number of **protons**, and therefore, typically identical to the number of **electrons**. The periodic table shows that element characteristics vary in a periodic, or predictable, way. The vertical columns are called **families** or **groups**. All elements in a group have similar chemical properties. The elements in each group have the same number of electrons in the outer orbital. Those outer electrons are also called valence electrons. They are the electrons involved in chemical bonds with other elements. Some elements have names such as **alkali metals** and **noble gases**. The other decimal number in the periodic table is the **atomic mass**, the average mass of a single atom of that element.

A nice copy of the Periodic Table can be printed from Jefferson Labs (part of the US Dept. of Energy): http://education.jlab.org/itselemental/

There are only **118 elements**, but we know that there are more than 118 substances in the world. This is because **atoms** chemically react with one another. When two or more atoms chemically bond to one another, a new substance with completely different properties can be made. The smallest unit of these new compounds is a **molecule**.

To make molecules, atoms must **bond** with one another. They do this in two different ways, both the result of the valence electrons from one atom interacting with the valence electrons from another atom. They can either trade atoms (**ionic bonding**) or share atoms (**covalent bonding**). Our activity today deals with covalent bonding. The atom cards are premarked with the number of bonds available.

Supplies

- **Atom cards**—print from Appendix, one sheet for every two to three students in class
- **Molecule cards**—print from Appendix, one set per student in largest class

Teacher Prep

- Print and cut out atom and molecule cards. Print on cardstock for durability.

Week 8 - Chemistry

Procedure

- "Do you remember watching the colored water mix by itself last week? This was because individual molecules are always moving. This week we will build molecules from atoms."
- Spread atom cards on table.
- Remember to prevent chaos by reinforcing self-control.
- Having students recite the science class rules is a fun recall exercise that should be done as often as possible.
- Explain that these atoms will combine in different ways to make different substances.
- Demonstrate a water molecule, showing how the bonds line up.
- "You may touch now; try to find the right atoms to make a water molecule."
- Give everyone time to make a water molecule.
- "Good! Now, we will make some more complex molecules."
- Pass out one molecule card to each student, starting with the simplest.
- Demonstrate how the bonds work on our atom cards.
- Allow them to begin to collect the atoms they need to make their molecule.
- They can raise their hand to show you their completed molecule. Praise them for a job well done.
- Repeat with more complicated molecules for as long as time allows. Come back together for the last five minutes to review. Molecules are the smallest unit of a compound. Molecules are made of atoms. The same atoms can combine in different ways to make different molecules.

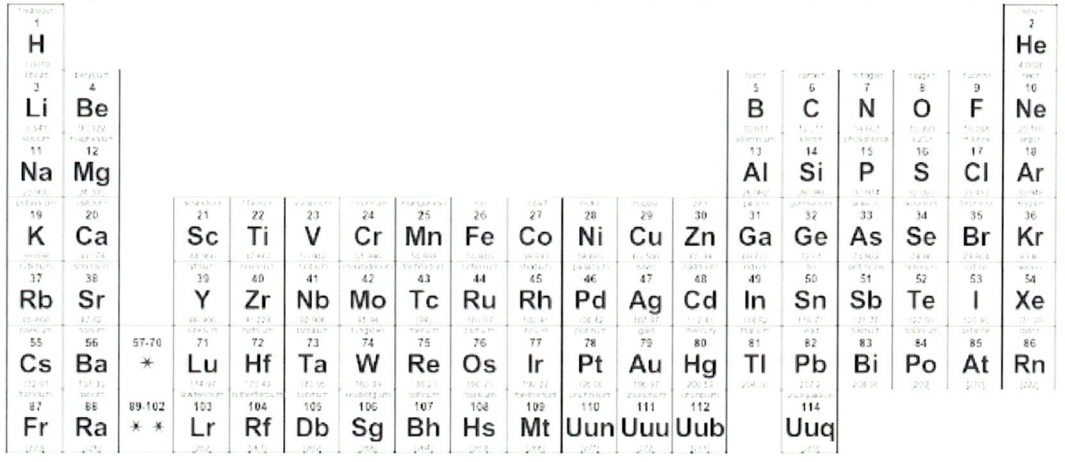

Bonus

Play the Periodic Table printable board game found here:
http://www.ellenjmchenry.com/homeschool-freedownloads/chemistry-games/documents/PeriodicTableGameUpdate.pdf

---

Science for Thought:

What is the smallest particle a piece of matter can be broken into without changing the type of substance it is?

That is, how small a particle of water is still water?

What about bread?

What about chocolate milk?

Chemistry - Week 8

Bouncy, Trouncy Polymer
Week 9

*A chemical reaction occurs when
bonds between atoms are created or destroyed.
Covalent bonds share electrons;
ionic bonds result when opposite charges attract.*

What is a chemical reaction?
What is a polymer?
What is the difference between an endothermic and an exothermic reaction?

Introduction

In chemistry, a **reaction** happens when two or more atoms or molecules interact. As the bonds of one molecule are broken, the atoms can recombine by bonding with other molecules. Sometimes they will create a completely new substance; sometimes, they produce heat, cold, or light.

To make molecules, atoms must **bond** with one another. They do this in two different ways, both the result of the electrons from one atom interacting with the electrons from another atom. They can either trade atoms (**ionic bonding**) or share atoms (**covalent bonding**).

For example, when iron molecules are exposed to oxygen in the atmosphere, a new substance, **rust**, is created. The presence of moisture speeds up rust because of the abundant oxygen in water. Copper molecules react with oxygen to make **copper oxide**, the greenish color we see on statues. Batteries use a chemical reaction. In an alkaline battery, the zinc reacts with oxygen, using ions from the potassium hydroxide electrolyte to create a new compound and release energy. During **photosynthesis**, which we studied in week 2, a chemical reaction occurs in the leaves of plants:
carbon dioxide + water + light energy → carbohydrate + oxygen + water

Supplies

- **Candle** and **match** or lighter
- **Chemically activated ice pack**—sold in first-aid section
- **Borax**—½ teaspoon per student
- **White Glue**—⅛ cup (30 ml) per student
- **Water**—80 ml per student (⅓ cup per student)
- Food coloring (optional)
- **Zip-top bags**—sandwich bags work fine, one per student
- **Graduated cylinders**
- ½ teaspoon measuring **spoons**—two per class is fine
- **Stir sticks**—popsicle, coffee straws, or plastic spoons
- **Paper cups**—one per student

Teacher Prep

- Pour borax in a bowl for easier distribution.
- Place ⅛ cup of white glue in zip-top bag for each student.
- Have container of water available—pour 80ml into a paper cup for each student.
- Set out graduated cylinders, water cup, and stir stick for each student.

Procedure—Part 1
- "Welcome, chemists." Explain what a **chemical reaction** is and the difference between **exothermic** (releasing heat, feels hot) and **endothermic** (absorbing heat, feels cold) reactions.
- Light a candle. "This is a chemical reaction. The wax and oxygen react to change into carbon dioxide, carbon (black stuff), water, and other things."
- "Is it endothermic or exothermic?" (exothermic)
- "Now, let's look at another reaction."
- Allow students to feel unbroken ice pack.
- Explain that a chemical reaction occurs when the inner package of the ice pack is broken, allowing the chemicals to mix.
- Break ice pack, and allow the students to feel what happens. (It gets cold.)
- Ask, "Was this an endothermic or exothermic reaction?" (endothermic)

Procedure—Part 2
- We will watch a very interesting chemical reaction occur when we mix some common substances together, making a polymer. A **polymer** is a long molecule, or chain of repeating, smaller units, held together by covalent bonds.
- Ask students if they remember the correct way to measure with graduated cylinders (eye level, measure at the lowest part of the meniscus).
- Pass out graduated cylinders and cups of water, zip-top bag with glue, and stir stick to each student.
- Instruct them to begin by measuring 30 ml of water. When you have ensured their accuracy, they can add the water to zip-top bag, seal, and mix.
- We will now add ½ teaspoon borax to the remaining water in the student's cup. (Have a parent helper go around the room with the bowl of borax and measuring spoon.)
- Stir your borax solution with your stir stick.
- Have students measure 30 ml of their new borax solution. If you wish to add color, add it here. One drop is plenty. (This will stain their hands, so I usually pass on the color.) Add contents of graduated cylinder to the zip-top bag; knead.
- The borax acts as the cross-linking agent or "connector" for the glue (polyvinyl acetate) molecules. Once the glue molecules join to form even larger polymer molecules, you get the thickened gel you see.

- Once the solution is mixed and solidified, students may remove their new polymer from the bag to play.
- If time allows, students can brainstorm fun names for their polymer.
- Return slime to bag for storage. It will last quite a while if stored in the refrigerator. If it molds or dries out, throw it away, and make more!

Bonus
Mixing baking soda and vinegar is a great way to witness a dramatic chemical reaction. Adding a few drops of dish soap will slow down the reaction and add drama. Remember to use large enough containers to contain the mess!

Did you Know?

Fire flies, or lightning bugs, make light through a chemical reaction.

The special cells in their abdomen contain a chemical called luciferin. *To make light*, the luciferin, after undergoing some changes, combines with oxygen to form a molecule called oxyluciferin, AMP, and light.

Purple Cabbage Juice
Week 10

Acids donate a hydrogen ion, while bases accept a hydrogen ion.
Acids have a pH of less than 7; bases have a pH of greater than 7.

How can we tell if something is an acid or a base?
If a substance has a pH of 4, is it an acid or a base?
Name three common acids.

Introduction

Acids and **bases** are all around us. Even things that don't seem like it are either mildly acidic or mildly basic. The exception to this is distilled water. The positive and negative ions in distilled water cancel out each other, making a **neutral substance**. The **pH scale** measures how acidic or basic a substance is. The scale deals with the concentrations of hydrogen ions (H+) and hydroxide ions (OH-).

Litmus paper, a paper scientist use, indicates a solution's **pH**. When these special strips of paper are dipped in a liquid, they change color to indicate a liquid's pH.

Substances with a **pH of less than 7 are acidic**. Acids taste sour. (Of course, we *never* taste things in science class.) Citric acid found in fruits is an acid. It turns litmus paper red and purple cabbage juice red. Substances with a pH **greater than 7 are basic**. Bases feel slimy and slippery and taste bitter. Soap is a base; as you know, it is slippery and tastes bitter. It turns litmus paper blue and purple cabbage juice blue. **Distilled water has a pH of exactly 7.**

Supplies

- **Purple cabbage juice** (made from red or purple cabbage)
- **Lemon juice**—20ml per student
- **Cheap shampoo (the most alkaline)**—20ml per student
- **Baking soda**—1-pound box per 30 students (3 teaspoons per student)
- **Medicine (eye) droppers**—one per student
- **Graduated cylinders**—one per student
- **Florence flasks** or pint Mason jars for mixing—two per student (plastic cups will work in a pinch)
- **Plastic teaspoons**—two per student
- Litmus paper (optional)—two per student
- **Paper towels**
- **Small paper cups** - three per student (lemon juice, shampoo, baking soda)

Teacher Prep

- To make purple cabbage juice, coarsely chop and boil red cabbage for 10—15 minutes, or until water is a nice purple color. Cool completely before using. This can be done up to a week ahead of time; store in the refrigerator.
- Because some shampoos are pH-balanced, test the shampoo to make sure it's basic.

Procedure

- We will use our graduated cylinders today — review how to measure using the bottom of the meniscus.
- Have students use medicine dropper to measure 10 ml of lemon juice with their graduated cylinder. Pour into flask or jar. Add a dropper of cabbage juice; swirl to mix.
- "What color is it?" (red)
- Measure 10 ml of shampoo. Pour into second container. Add dropper of cabbage juice.
- "What color is it?" (blue; if it is not blue, it's possible that your shampoo is pH-balanced.)
- Have students measure 5 ml of shampoo and add it to the lemon juice. Swirl. Was it enough to turn the mixture basic? Continue adding 5 ml at a time until it has changed from an acid to a base.
- At this point, introducing baking soda (a base) is lots of fun. Allow students to gradually add one teaspoon at a time to either container. This might get messy!
- Students can continue to experiment and have fun. The cabbage juice changes colors when you change the pH of the substance. Add more if necessary.
- The pictures below show the range of colors possible—more acidic on the left, basic on the right.
- Pass out a piece of litmus paper to each student. Teach them to dip, then remove it from the substance. Compare the resulting color to the chart that came with your litmus paper to determine the pH. Litmus paper is not as dramatic or fun as the cabbage juice, but it is still a good scientific tool to which to expose students.

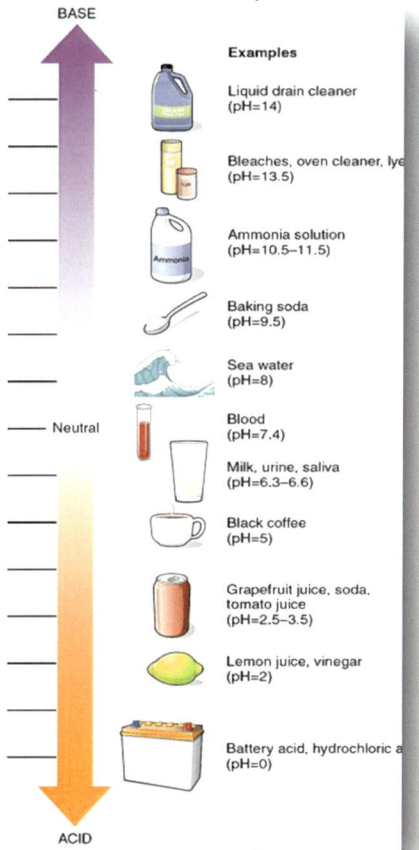

Results of adding cabbage juice to solutions of different pH values.

Bonus

If you have old coins around the house that you wish to make part of a collection, you can clean them easily. Ketchup gives excellent results; just let stand for a few hours. Test the ketchup with your cabbage-juice indicator to find out why this works.

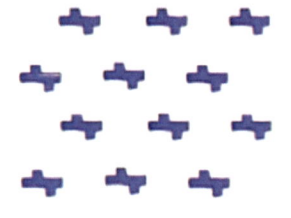
Homogeneous

Heterogeneous

Homogenous or Heterogeneous?
Week 11

In chemistry, there are two types of mixtures: homogeneous, which means the same throughout, and heterogeneous, not the same throughout.

What is a heterogeneous mixture?
What is another name for a homogeneous mixture?
What is the purpose of chromatography?

Introduction

When substances are mixed together, scientists call this a **mixture**. Mixtures can be either heterogeneous or homogeneous. **Heterogeneous mixtures** contain distinguishable particles. They can usually be easily separated by physical means such as **filtering**, **evaporating**, or **decanting**. Examples of heterogeneous mixtures are sand, orange juice with pulp, pizza, and concrete.

Homogeneous mixtures are also known as solutions. They cannot be separated by physical methods. The smallest sample of a solution retains the characteristics of the original. Examples of solutions are air, dishwashing detergent, and metal alloys.

Today's experiment involves a process called chromatography. **Chromatography** (literally, to write with color) is a method of separating complex mixtures to analyze them. We will separate black ink, which, although it looks homogeneous, is a **suspension** of pigments in a carrier fluid. The various pigments in the black ink will migrate through the filter at different rates. This differential rate of migration makes chromatography valuable to industry. It can be used to analyze a substance or to separate fragile components for further use.

Supplies

- **Mixture of house items in large bowl** to demonstrate heterogeneous mixtures (rice, beans, coins, Legos, cotton balls, and so on)
- **Coffee filters**—one per student
- **Glass pint jars**—one per student in largest class
- **Black Crayola marker or similar washable markers**—*Sharpies do not work*. Two to three per class.
- **Scissors**—two to three pair per class should be plenty.
- **Water**—one cup per student
- **Paper towels**—one per student

Teacher Prep

- Test to make sure that the chromatography works with your markers!

Procedure—Part 1

- After explaining the difference between heterogeneous and homogeneous, allow students to look at and touch the large bowl of mixed items.
- Ask, "Do you think this mixture is heterogeneous or homogenous?" "How can you tell?" (Heterogeneous, we can see the different particles.)
- Show students the tip of a marker.
- "Do you think the ink is a heterogeneous mixture or a homogeneous solution?"
- Review characteristic and definition. (It looks the same throughout.)
- Let's do an experiment to find out.

Procedure—Part 2

- Pass out coffee filters and markers.
- Use scissors to cut a small (¼") cross in the center of each filter.
- Use markers to draw a dashed circle about the size of a quarter around the cross.
- Pass out glass jars, fill halfway with water.
- Pass out paper towels. Roll into a cylinder and poke through the hole in your coffee filter.
- Place coffee filter over your jar of water, with paper towel reaching into the water. Coffee filter itself should not touch the water. (This allows the water to wick up the paper towel into the coffee filter.)
- Watch, as wicking action brings the water up to the coffee filter and disperses the ink.
- Beautiful patterns should occur.
- What we have just done is called *chromatography*. It is a process used to separate substances into their parts. What a surprise! Black ink is actually a mixture of different pigments suspended in a carrier! It is not a solution at all. It is a heterogeneous mixture; the parts were just too small to see.

Research colloids, which are another type of mixture. Try to label all the substances in your refrigerator as a heterogeneous mixture, solution, or colloid.

Did you Know?

The words *homogeneous* and *heterogeneous* come from the Greeks.

•

The prefix *homo-* indicates sameness.

•

The prefix *hetero-* indicates difference.

Chemistry - Week 11

Exploring Density
Week 12

Four states of matter are solids, liquids, gasses, and plasma.

What are the three chief states of matter?
What causes a change of state?
What is density?

Introduction

Remember the atoms we talked about in weeks 7 and 8? We can't see them, but these atoms always move. For example, water, or H_2O, is always the same substance, regardless of how fast the atoms move. When they move very slowly, they can fit more closely together—we call this ice. When they move a little faster, they need more room. Because there is more space among the molecules, they can move, or flow. We call this water. When the atoms move very quickly, they bounce off one another and don't even stay in a container. We call this water vapor. Ice is the **solid**, water the **liquid**, and water vapor the **gas** phase of water.

All substances have these three states of matter. Some can also be found as a super-charged gas, known as **plasma.** To change matter from one state to another, we must add or remove energy. If we heat ice, it melts. That is adding energy. If we cool water, it freezes. That is removing energy. If we add energy to a gas, it becomes plasma.

As of 1995, scientists have identified five states of matter. In addition to solids, liquids, gasses and plasma, there is a new one called a **Bose-Einstein condensate.** The first four we have known about for a while. The scientists who worked with the Bose-Einstein condensate received a Nobel Prize in 1995.

The word for how tightly packed atoms are in a substance is **density**. Density is the amount of matter in an object. Liquids are denser than gases; solids are generally denser than liquids. Substances have different densities. Less dense substances rise above or float on denser substances.

$$\text{Density} = \frac{\text{Mass}}{\text{Volume}}$$

Supplies

- **Tall, clear container**—carafe, vase or quart mason jar—one per class demo
- **Heavy solids**—washers, pebbles, marbles…
- **Small piece of wood**—make sure it floats
- **Water**
- **Ice cubes**
- **1 pint mason jar**—for corn oil/ice cube demonstration
- **Graduated cylinders**—one per student in class
- **Corn oil**—⅛ cup per student
- **Corn syrup**—⅛ cup per student
- **Water**—⅛ cup per student
- **Paper cups**—three per student to distribute oil, corn syrup, and water

Procedure—Part 1

Density is an important concept in chemistry. Today's first activity will be a class demonstration that explores the concept of density.

- Pour water into a tall clear container.
- Drop a heavy, solid object into the water. "What happens?" (It sinks.)
- "Why?" (Many solids have higher densities than liquid does.)
- Try other objects, if desired.
- Show solid wood to students. "Is this a solid?" "Let's see if it is denser than water is."
- Drop wood in the water. "Does it sink?" "Is it denser or less dense than water is?" (less dense) This block of wood has less matter in it than an equal volume of water has.
- "Did you notice that the water level went up when you placed the first (heavy) object in the water?" (The amount the water level increases is one way to measure an object's volume. With the volume and the mass, we can calculate its density.)
- Fill a pint Mason jar with corn oil. Drop an ice cube in the corn oil. Observe that it floats, although not necessarily at the top. Set aside for later.
  (Check on it periodically to catch it before ice melts completely. When you notice that the ice is about half-melted, carry jar around for students to observe. The melted water will have sunk to the bottom of the jar, while the rest of the ice still floats in the corn oil, demonstrating the different densities.)

Procedure—Part 2

- Next, let's explore whether liquids can have different densities.
- Pass out graduated cylinders.
- Give each student a small paper cup of corn syrup.
- Measure approximately 10 ml of corn syrup into your graduated cylinder. "Did it sink below the air?" (Yes, it is denser than air is.)
- Pass out small paper cup of cooking oil.
- Have students gently add 10 ml of oil to their graduated cylinder so it doesn't mix. "Which is denser?" (If it mixes, it will take time to separate.)
- Pass out small paper cup of water.
- Have students add 10 ml of water to their graduated cylinder. Watch as the water settles between the corn oil and the corn syrup.
- "Which of the three liquids was the densest?" "Least dense?" (The one on top is the least dense.)
- The layers are formed because of density
- .If you haven't already done so, look at the corn oil and ice you set aside.

Bonus

If dry ice is available locally, explore sublimation. Add a little Dawn dish soap to a large bowl of water. Drop dry ice in bottom (Do NOT allow students to touch dry ice.) When bubbles rise to the surface, they will be filled with a cloudy gas that is pure carbon dioxide. Bubbles are safe to hold. When students pop them, the cloudy carbon dioxide will float off into the air.

Did you Know?

Most solids are denser than their liquid counterpart. One exception to this is water. At 39 degrees Fahrenheit, water begins to expand as it cools. As a result, ice is less dense than water.

This miracle of God's creation allows ice to float on lakes and the ocean. Without this miracle, life in the oceans would not be possible!

Chemistry - Week 12

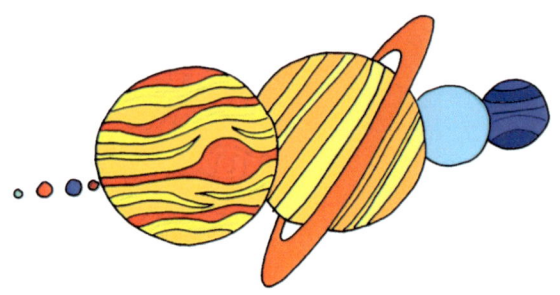

Walking the Solar System
Week 13

There are eight planets in our solar system:
Mercury, Venus, Earth, Mars, Jupiter, Saturn, Uranus, and, Neptune

Can you recite the order of the planets?
Which is the largest planet?
Which is the smallest?

Introduction

The **solar system** is huge! We live on a big planet, but compared to the solar system, it is tiny. Our activity today gives the students an idea of just how huge it is.

Our **Sun** and **planets** make up our solar system. The solar system gets its name from the Greek word *sol*, meaning "sun." Planets vary in size from the tiny Mercury to the largest gas giant, Jupiter. Although some of us learned that the ninth planet is Pluto, scientists now consider Pluto a space object, not a planet. Reasons for this change were the new information that Pluto is irregular in shape and also has an irregular orbit.

If our solar system is this large compared to us, just think how amazing our infinite God is!

Supplies

- Sun—a **balloon** that can be blown up to an 8-inch diameter **or yellow ball**
- Mercury—a candy **sprinkle** used to decorate cakes, 0.03-inch diameter
- Venus—a **peppercorn**, 0.08-inch diameter
- Earth—a second **peppercorn**
- Mars—a second **sprinkle**
- Jupiter—a **pecan, walnut, or chestnut,** 0.9-inch diameter
- Saturn—a **filbert or acorn,** 0.7-inch diameter
- Uranus—a **peanut or coffee bean**, 0.3-inch diameter
- Neptune—a second **peanut or coffee bean**
- Pluto—a third **sprinkle**

Teacher Prep

Today's class will meet outside. Find a place in advance from which you can walk 1,000 yards (3/10 mile) in somewhat of a straight line.

If this is not possible, or weather is a problem, use the objects indoors. Use a gym and smaller, 1-foot paces instead of long paces. Or, have extra objects on hand to give each student a set, then have them arrange the planets in order. Spread them across the room in an approximate scale model using inches instead of paces if you have 20 feet available. If not, use centimeters.

Week 13 - Astronomy

Procedure

Thanks to Guy Ottewell for the following activity. It is used with his permission. Please visit his website: www.universalworkshop.com/TYM.htm

- Pass out peppercorns for each student to hold so they may visualize the size of the earth in our model.
- In our activity today, these represent the Earth on which we live.
- The earth is 8,000 miles wide. The peppercorn is 0.08 of an inch wide.
- Hold up the sun; notice how much bigger the sun is than the earth.
- In our model, 1 inch = 100,000 miles.
- This means that 1 yard (36 inches) represents 3,600,000 miles.
- Take a giant step.
- Tell the students that on our scale model, this distance, one pace across the floor, is an enormous space journey of about 3,600,000 miles.
- Ask the students: "What is the distance between the earth and the sun?"
- It is 93 million miles. In our model, that is 26 yards, giant steps, or paces.
- This still might not mean much until someone starts to measure 26 paces.
- Hand the sun and the planets to members of the class. Be sure they know the planet they are carrying and can produce it when called.
- Put the sun down and march away as follows. (Appoint someone else to be your "Spacecraft" or "Pacecraft"—so that you are free to talk.)
- Take 10 paces. Call for Mercury and have the Mercury bearer put down his or her planet (sprinkle).
- Another 9 paces. "Venus—put down a peppercorn."
- Another 7 paces. "Earth."
- Already, the solar system seems unbelievable. Mercury is supposed to be so close to the sun that it is merely a scorched rock. As for the earth, who can believe that the sun could warm us if we are that far from it?
- Another 14 paces. "Mars." Now, come the gasps at the first substantially large leap.
- Another 95 paces to Jupiter.
- Here is the "giant planet," a mere nut and more than a city block from its nearest neighbor in space. From now on, amazement itself cannot keep pace, as the intervals grow extravagantly:
- Another 112 paces. "Saturn."
- Another 249 paces. "Uranus."
- Another 281 paces. "Neptune."

You have marched the entire solar system. Time to look back and observe the size and distance of your planets!

Bonus

Build models of the planets to create your own solar system. The planets can be researched and drawn with the correct size and colors, or printed from the Internet. Hang it from the ceiling to give your home an "out-of-this-world" feeling!

Did you Know?

Betelgeuse is one of the largest stars in the universe, with an average diameter of 5.5 AU!

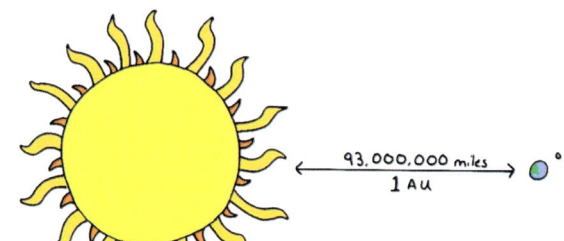

How Hot is the Sun?
Week 14

The distance from the sun to the earth is 93 million miles. Astronomers call that distance an astronomical unit.

How far is the earth from the sun?
What are the two most abundant elements in the sun?
How many AU is it to Pluto? To the nearest star?

Introduction

Distances in space are huge! If we tried to measure in earth-sized measurements such as feet or miles, the numbers would be very large. Scientists make this easier by using the **astronomical unit (AU)**. Even with the AU, distances in space are difficult to imagine. **Pluto is 39 AU** from the sun on average. **Alpha Centauri**, the nearest star to our solar system, is about **271,000 AU** from the sun! The distance from the sun to the center of the **Milky Way Galaxy is about 1,700,000,000 AU!**

The **sun is about 75% hydrogen and 24% helium**. Much of the matter in the sun is in a state of super-heated gas, known as **plasma**. Through **nuclear fusion,** the plasma "burns" to create vast amounts of energy. As the hot plasma rises to the surface, lower temperature plasma falls, creating currents of circular motion known as convection. All this movement of particles creates great magnetic fields within the sun. **Solar flares** can be seen to arc along the patterns of these **magnetic fields**. The plasma that reaches the surface can cast off particles that move through the solar system creating what is known as the **solar wind**. Imagine, right now, particles are traveling from the sun toward you! Most pass through objects unnoticed, whereas some interact with the earth's magnetic field creating the beautiful phenomenon known as **aurora borealis,** or northern lights.

Supplies

- **Colored paper** (and labels for the youngest children) as follows:
  Core, 27,000,000 °F, **white** (3" circle)
  Radiative zone, unknown, **pale yellow** (4.5" circle)
  Convective zone, varies, **yellow** (6" circle)
  Chromosphere, 6,000-500,000 °F, pale **yellow** (8.5" circle)
  Photosphere, 10,000 °F, **dark yellow/gold** (7.5" circle)
  Sunspots, 4,000 °F, **brown** (random blobs)
  Solar flares, **orange** (hollow loops, cut 2" long to have 1" gluing surface, 1" extension)
  Corona, 2,000,000 °F, **white** (white branching, feathery shapes, up to 5" long)
- **Brass fasteners**—one per student
- **Scissors**—one per student in class
- **Glue sticks** and **preprinted labels with above zones** OR
  Pencils for writing zones on the sun

Teacher Prep

Cut out circles as above. Students may cut their own sunspots and solar flares. Print labels for your nonwriters. You might wish to have enough for your beginning writers as well to speed the project.

Procedure

- Take a few moments to talk about the distances experienced in last week's activity. Translate these into thinking about distances in space using the AU. Move to the activity quickly, as it takes some time. You can talk more about distances at the end, if time remains.
- Describe the sun. Review the states of matter from week 12. "The sun is an example of plasma in the universe."
- "What part of the sun do you think is the hottest? The coolest? Our model of the sun will give us these answers."
- Pass out *core*. Poke hole in center and poke brass fastener through from the top. Attach label (youngest students), or write "Core 27,000,000 °F." Teach students about the degree symbol and F = Fahrenheit, if necessary. This is the hottest part of the sun.
- Pass out *radiative zone*. In this zone, solar material is dense enough to transfer the enormous heat from the core through radiation (not radioactivity), just like a radiator in a house or a standard oven. Poke hole and add it to the back of the core. Attach or write label as close to the outer edge as possible.
- Pass out the *convective zone*. Here, the plasma's density is not great enough to transfer heat through radiation. Instead, great movement known as convection currents transfers the heat outward. A pot on the stove has **convection currents**, as does a convection oven (caused by the fan). Because of this convection, there is great variance in temperature. Poke hole, and add it to the back of the radiative zone. Attach or write label as close to the outer edge as possible.
- Next, we will assemble the *photosphere,* or visible surface of the sun. We use darker colors to represent cooler temperatures. First, poke a hole in the circle. Next, cut a notch about ⅓ of the circle to allow students to see the inside layers. Be careful NOT to cut all the way to the center hole.
- Remove brass fastener to put photosphere on the top. Label photosphere.
- Allow students to cut random brown blobs for *sunspots*; glue around surface. Label sunspots.
- Allow students to cut hollow loops for *solar flares*. Glue to back of photosphere so 1" of loop is exposed around edge.
- Pass out *chromosphere*. This is part of the solar atmosphere, and it is named for the flash of color seen briefly before and after a total solar eclipse. The coolest area of the sun is here, but the temperature rises as distance from the sun increases. Label and attach to back of convective zone.
- Pass out white paper to make the *corona.* The corona continuously expands into space, creating the solar wind. It extends the entire distance of our solar system! Notice that it is the second hottest temperature observed near the sun. This is the part visible around the earth in pictures of a solar eclipse. (Never look at the sun, especially during an eclipse, as it will damage your eyes.) Cut spikes and points as desired into the white paper. Glue to back of the chromosphere so they protrude. Label the corona.

Bonus

Make a sun pinhole viewer. It is the only safe way to view the sun without special equipment. Directions can be found online to make this from a basic oatmeal box and black paper. A #10 welding helmet also offers protection to view the sun.

Did you Know?

It takes about 8 minutes for light from the sun to reach the earth!

Seasons
Week 15

*The earth revolves around the sun.
This is called its orbit.
The earth rotates on its axis,
giving us night and day.*

How many rotations are in a day?
Can you describe the rotation of the earth?
How long does it take the earth to make one revolution around the sun?

Introduction

The earth **revolves** around the sun, making one revolution every 365¼ days. (The extra quarter turn results in a **leap-year** day every four years, which we place on February 29.)

Because the earth is **tilted**, rather than completely vertical, on its **axis**, we have seasons. The Northern Hemisphere is tilted away from the sun during the winter, making the angle that the sun hits the earth less direct and resulting in cooler temperatures. In the summer, the Northern Hemisphere is tilted toward the sun. The light from the sun hits at a more direct angle, making the earth warmer.

While the earth revolves, it is also **rotates**, or spins, on its axis. One rotation is what we call a *day*. The day is divided into 24 hours. As the earth rotates, the sun seems to rise in the east, moving overhead as the day passes, and then setting in the west. This is an illusion; it is the earth that moves! At the equator, the days and nights are each 12 hours long. In the summer, as you move north, the days get longer and the nights shorter. The opposite is true in the winter when northern days get shorter. North of the **Artic Circle,** the days are so short that the sun never rises during the winter!

The **Tropic of Cancer** line on your globe represents the northernmost travel of the sun's direct rays, which occurs on the **Summer Solstice**, June 20, 21, or 22. North of this line, the sun is never straight overhead.

Supplies

- **Globe** that can be carried by students
- **Lamp** with shade removed

Teacher Prep

- Set up lamp in the center of room.
- Make space so students can move freely around lamp.

Procedure

Students who did the solar energy activity in the electricity unit in year 1 will remember that more energy was available when solar cells were perpendicular to the light rays; less when the rays hit at an angle. Today's activity demonstrates how the angle of the sun determines the seasons.

- Assign a student to be the sun. He or she should stand in the center of your area.
- Assign a second student to be the earth. He or she should hold the globe and stand about 3 feet from the sun. "Is it day or night in North America?"
- Have student position the globe to its proper tilt. "Where is it the hottest now?" (Where the rays are the most direct.)
- "Which hemisphere is the sun shining most directly on? It is summer here." 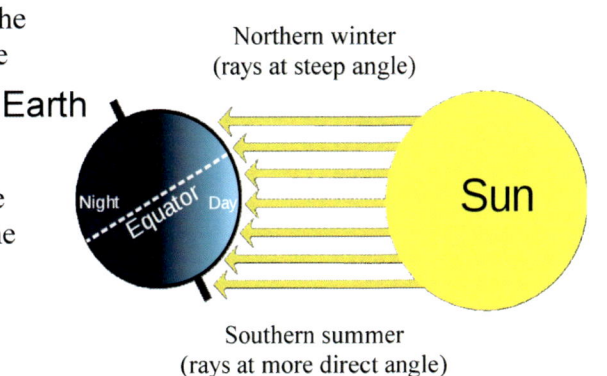
- "Where are the sun's rays the least direct? It is winter here."
- Have student walk in a circle to the opposite side of the sun. Which hemisphere is now the hottest? The coolest?
- Trade jobs so another student has a chance—repeat for spring and summer.
- If time allows, have each student walk a complete revolution while holding the globe.
- Extra time can be used reinforcing rotation compared with revolution. Put globe down, and have all students form a circle around the sun. Begin to walk in a circle as a group. This is revolving.
- Have students stop in place and begin to spin. This is rotating.
- Now, challenge students to revolve and rotate at the same time. This is what our sun does at exactly the correct speed for us to have seasons. Isn't our Creator magnificent?

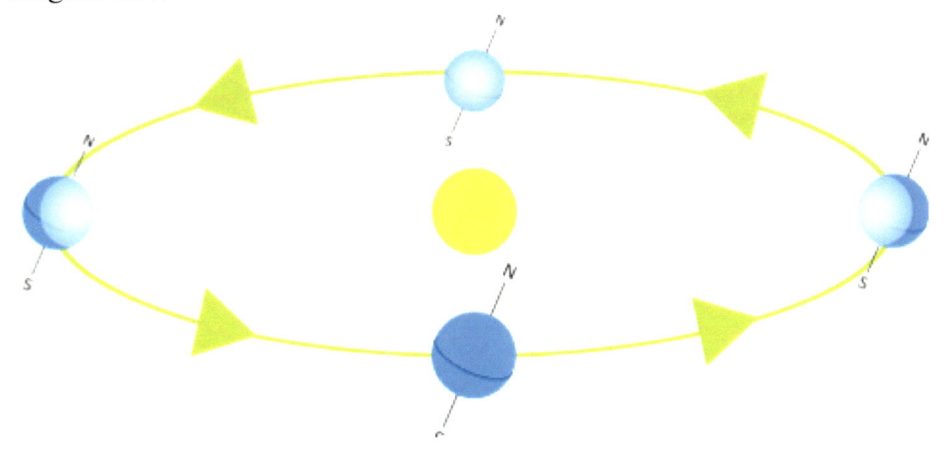

Bonus

The earth is really in orbit around the sun. Find out what other objects orbit the earth—there are more than just the moon! You might come across the term *geosynchronous orbit*. Can you explain what this is?

Autumn Fires

In the other gardens
And all up the vale,
From the autumn bonfires
See the smoke trail!

Pleasant summer over
And all the summer flowers,
The red fire blazes,
The grey smoke towers.

Sing a song of seasons!
Something bright in all!
Flowers in the summer,
Fires in the fall!

Robert Louis Stevenson

Astronomy - Week 15

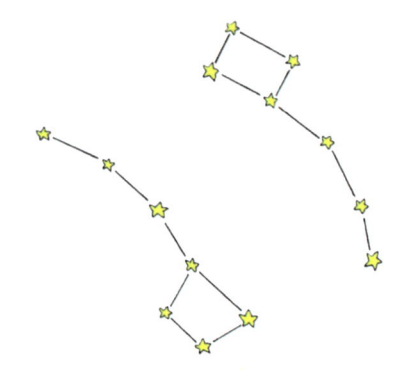

Patterns in the Sky
Week 16

Stars are balls of hot gas, mostly hydrogen and helium.
Polaris is the North Star. Sirius is the brightest star.
Alpha Centauri is the closest star besides our sun.

What is another name for Ursa Major?
How were constellations used by ancient people?
What is your favorite constellation?

Introduction

Constellations are like connect-the-dot pictures with stars.

Our current official list of 88 constellations has been in use since 1922. It is based on a list that goes back 2,000 years to the Greek astronomer **Ptolemy** (born in AD 90). Ptolemy was a Roman citizen of Greek ancestry who lived in Egypt in the second century AD. He is a noted astronomer and mathematician. Several of his books and charts survive today.

Constellations have been used to aid in navigation from the earliest times. Constellations also are a way of sharing stories and legends among people.

The Bear, Orion, and the Pleiades are mentioned in chapters 9 and 38 of the book of Job in the Bible.

The constellations we will work with today are the following:

Ursa Major, or the Great Bear; also known as the Big Dipper
Ursa Minor, or the Little Bear; also known as the Little Dipper
Draco, the Dragon
Cepheus, King of Ethiopia
Cassiopeia, the queen from Greek mythology

Supplies

- **Space Grid**—in Appendix, one per student
- **Coordinates of various constellations**—in Appendix, print and/or write on board.
- **Pencils**—one per student with erasers
- **Heavy black paper or cardstock**—one to two per student
- **Pins** to poke holes—one per student in class
- **Scotch tape**

Teacher Prep

- Plot the constellations for your youngest students before class, if desired.

Week 16 - Astronomy

Procedure—Part 1

With the youngest students, you might wish to skip to part 2 with preplotted constellations.

- A **coordinate grid** is a system in which points are represented by letters and numbers. Numbers represent the lines in one direction, letters represent those in the other direction. By giving letters and numbers, any point on the grid can be located. Maps often use this system.
- Demonstrate the system to students by plotting coordinates of a small sample constellation.
- Pass out Space Grid sheets and the coordinates for one constellation.
- You might wish to plot an easy one step-by-step together to show students the process.
- Give students ample time to plot an additional constellation, helping as needed.
- Have extra constellation coordinates ready for the students who finish quickly.

Procedure—Part 2

- When all students have made at least one constellation, pass out pieces of black paper.
- Ask students which constellation is their favorite. They will now make a viewer to display that constellation.
- Attach the black paper behind the grid sheet on which they plotted their constellations with a couple rolls of Scotch tape, being careful not to put the tape where there is a star.
- Use a pin to poke a neat hole through both pieces of paper at the location of each star.
- Hold up to a light to see your constellation.
- At home in a dark room, shining a flashlight behind the constellation will give a more dramatic effect.

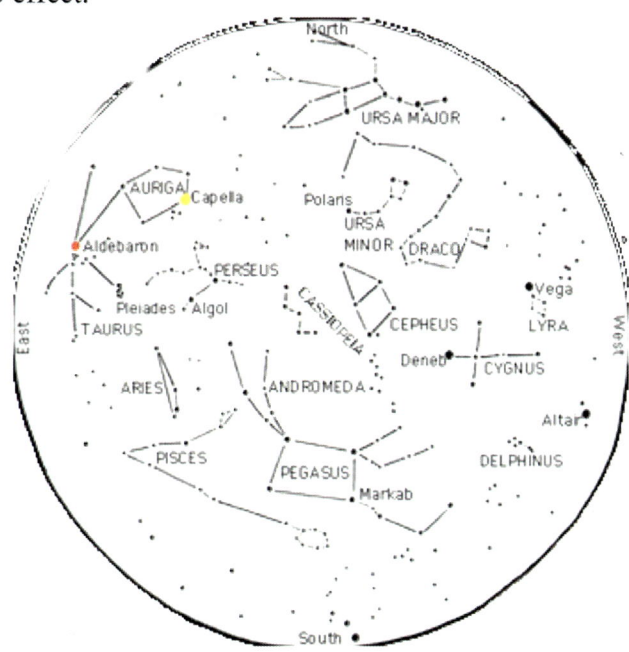

Bonus

Cut circles out of glow-in-the-dark paper and attach to bedroom ceiling to form constellations.

"When you reach for the stars, you may not quite get one, but you won't come up with a handful of mud either."

—Leo Burnett

Astronomy - Week 16

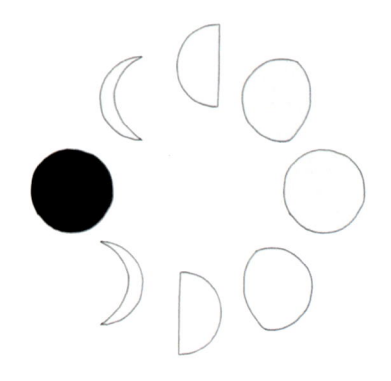

Building Rockets
Week 17

*The phases of the moon: new moon,
waxing crescent, first quarter, waxing gibbous, full moon,
waning gibbous, third quarter, waning crescent,
and then it starts again.*

Who was the first man in space?
What was the first animal in space?
Who was the first man on the moon?

Introduction

The race to space was an exciting period in history with many firsts.

The United States and the Soviet Union were competing against each other to see who could break free from Earth's gravity first. Many trials were run to test various aspects of space before sending a human up in a rocket.

The first animals sent into space were fruit flies launched in a V-2 rocket in 1947 from White Sands Missile Range in New Mexico. The fruit flies were recovered from the ejected capsule alive.

In the ensuing years, many other countries joined the space race sending a variety of animals into outer space, a experimentation process still in place today.

On April 12th, 1961, Yuri Gagarin, a Russian Soviet pilot and cosmonaut was the first human to journey into outer space. He completed an orbit of the Earth in his spacecraft Vostok.

It took eight more years before we successfully landed a man on the moon. July 20, 1969, with the Apollo 11 mission, Neil Armstrong was the first human to step on the moon. Armstrong became known for saying, "One small step for man, one giant leap for mankind." as he stepped on the lunar surface.

Supplies

- **Super 6 rocket kit**—one per student, plus a couple of extras from http://www.pratthobbies.com/
- **Rocket engines**—one per younger student; two per older student
- **Launch pad and ignition system**
- **White glue**
- **Paper towels**
- **Markers and highlighters** for decorating rockets

Teacher Prep

Build a rocket to have as a finished model to show students. Follow the instructions that come with the kit you have purchased to become familiar with the procedure.

Procedure

- Have the students focus on building their rockets today. If any finish, they can spend the rest of class decorating their rockets.
- Write names on rockets.
- Keep the rockets and manufacturers' launch instructions until week 18 when we launch our rocket. (If wind is a frequent problem in your area, consider attempting the launch in week 17 so you have another week available if you need to postpone the launch.)
- If any rockets are not finished, either take them home yourself to complete (recommended) or ensure that the parent understands the procedure and agrees to complete the work with his or her student before the next class.

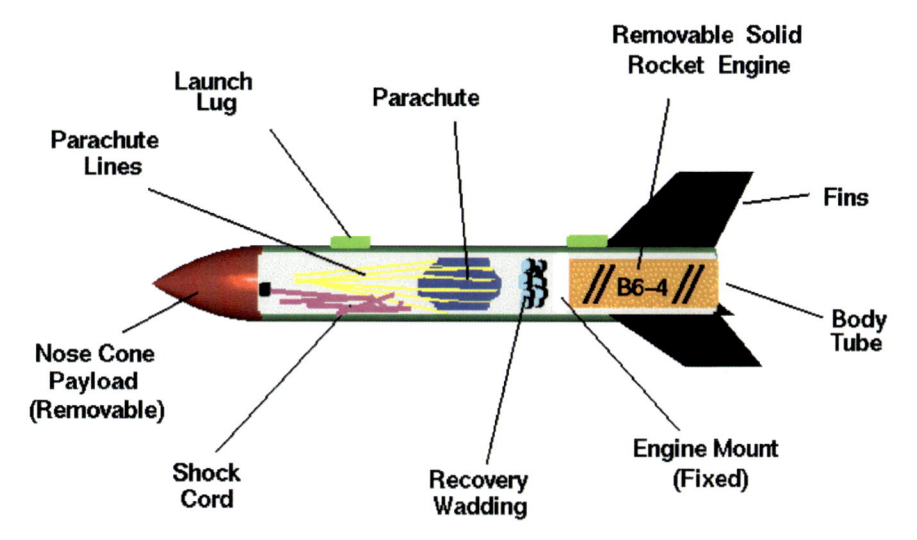

Image courtesy of NASA

While students are working on their rockets, share facts about the moon.

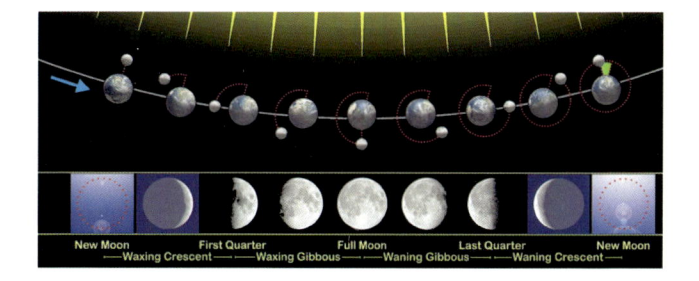

Bonus

Make a phases of the moon poster. Either gather pictures from the Internet, or for a truly involved activity, have students take a picture of the moon each night to use in their custom presentation.

"For every one, as I think, Must see that astronomy compels the soul to look upward and leads us from this world to another."

—Glaucon, Plato's older brother, in Plato's The Republic, c. 380 BCE

Rocket Launch
Week 18

Here are some other objects in space: comets, asteroids, black holes, nebula, and galaxies

What is the difference between a meteor and a meteoroid?
What is a supernova?
Can you see a black hole?

Introduction

Comets are space objects composed of ice, dust, and rocky debris. When a comet nucleus nears the sun, solar energy begins to heat the ice and vaporize it. The gases form a cloud called the **coma** around the nucleus. Some gas is stripped of electrons and blown back by the **solar wind**. The dust particles are pushed away from the comet by solar radiation, forming a dust tail that can be many millions of miles long.

Meteors, known also as *shooting stars*, are usually sized from a grain of sand to about a softball. As the meteor enters the earth's atmosphere, it becomes very hot because of **friction** between the meteor and the air molecules. The high speed results in so much heat that they become bright and seem to streak across the sky as they burn up. (You can feel heat from friction by rubbing your hands together quickly for a few moments.) When a large particle of a meteor actually hits the earth, instead of burning up, it is called a **meteoroid**.

Asteroids are the small rocky objects in the solar system. The largest asteroid is Ceres, which is 933 kilometers (580 miles) across. They do not have enough gravity to pull themselves into a sphere.

Black holes are very fascinating to study. They are regions in space where gravity is so strong that nothing can escape, not even light. The largest are thought to form when a star collapses. There are many other fascinating objects in space to study, including **supernovas** (images of exploding stars), **nebulae**, globular clusters, and so on.

Supplies

- **Rocket launching equipment**—pad and igniter—see week 17.
- **Rocket engines**—one to two per student
- Spare **batteries** if your igniter is battery-powered
- Students' completed **rockets**

Teacher Prep

Find an open area outdoors, as free of trees as possible, to launch your rockets. A playground or parking lot is usually perfect. Be advised that launching in windy conditions might be tricky.

Week 18 - Astronomy

Procedure

- Position your launching equipment to allow for the wind. If your rockets don't get lost in trees, they can be saved and reused.
- Students should stand at a safe distance of at least 20 feet from the launch pad while their classmate is launching.
- Position rocket on launcher; follow directions with your launching system.
- Students enjoy doing a countdown with each launch.
- Make sure you read and follow all safety procedures given by the manufacturer of your rocket.
- Launch away!
- This concludes our study of space. You could offer to bring the launching equipment to any social gatherings you might have in the future if the students would like to launch their rockets again.

Image courtesy of NASA

Did you Know?

To exit the earth's gravity, the space shuttles must have 6.8 million pounds of force!

The name comet comes from the Greek word meaning "longhaired" because of its tail.

Bonus

Make a chronological list of all the US space missions. Compare this to a list of space missions from another country, such as Russia, Japan, or China. Some interesting things can be observed.

Astronomy - Week 18

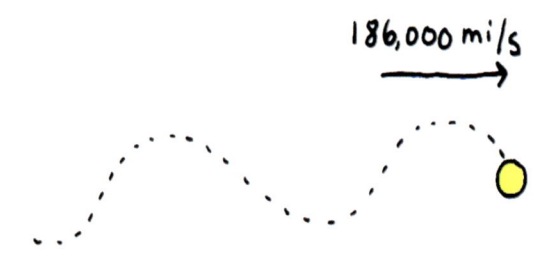

Polarizing Light
Week 19

Light has properties of both waves and particles called protons. The speed of light is 186,000 miles per second.

What is the speed of light?
Is light a wave or a particle?
What does it mean to polarize light?

Introduction

Light is **energy** that our sight can detect. It is made of electromagnetic radiation and travels in a straight path. At **186,000 miles per second**, light is faster than anything. In fact, according to Einstein's theory of relativity, it is impossible for anything to surpass the speed of light. We use a **light-year**, or the speed light travels in one year, as another way to measure distances in space.

We often hear about light waves and the wavelength of light, but for hundreds of years, scientists have argued whether light is really a wave. Back in the seventeenth century, the brilliant English scientist, **Sir Isaac Newton**—one of the first people to study the subject in detail—thought light was a stream of "corpuscles," or particles, but his rival, a Dutch scientist named Christian Huygens, insisted that light was made of waves. Now, we know that they were both right. Light has properties of both a **wave** and a **particle**. So which is it? You could say neither and both. It is simply energy that doesn't exactly match either.

Light waves come from a source in all directions—vertical, horizontal, diagonal. We can eliminate the waves in all directions but one resulting in **polarized light**, in which waves only travel in one direction. This is why better sunglasses are polarized. They don't just darken; they remove the glare that results from light waves reflecting off shiny surfaces. An interesting thing happens if you put light through two polarizing filters placed at a 90-degree angle; you will effectively filter, or remove, all the light!

Supplies

- **Polarizing light sheets**— two 2"—3" squares per student, cut from large sheets
- **Light source**—flashlight or lamp
- **2 ropes**—jump ropes work fine.
- **2 poles**—broom handles work fine.

Procedure—Part 1

- Discuss properties of light.
- Pass out one polarizing sheet to each student.
- Have them hold it up to the light. Do they notice any change in the light?
- Pass out a second polarizing sheet.
- Hold it over the first sheet.
- Do they notice any changes in the light? (The amount of change will vary depending on whether the two sheets' polarizing directions are perpendicular or not.)
- Have students turn the top sheet 90 degrees, or ¼ turn.
- Discuss what happens now.
- Bring out lamp, flashlight, or other direct light source.
- Allow students to view the light through one sheet and both sheets, rotating sheets to get different effects.

Procedure—Part 2

- Bring out jump rope.
- Have two students each take an end and move rope up and down to get a wave pattern.
- This simulates a one-directional wave.
- Bring out second rope.
- Have two students move it side to side to get a wave pattern.
- This also simulates a one-directional wave but in another direction.
- Now, hold the two poles vertically next to each other about 6" apart.
- Have the two sets of students try to recreate their wave with the rope between the poles.
- It will only be possible for the vertical wave. This is how the polarizing sheet works. It eliminates the waves in all directions but one.
- Imagine that if you held two more poles horizontally over the two vertical poles. It would make a little square, and no wave would be possible! This is what the polarizing sheets did when you aligned them at 90-degree angles—they blocked all the waves, and thus, all the light.

Did you Know?

Iceland Spar is a type of calcite crystal that polarizes light. It might be the "sunstone" used by the Vikings to navigate by the sun on cloudy days.

Bonus

A *radiometer* is a device that can demonstrate the particle component of light. One can be purchased from www.scienceonline.com. When a radiometer is placed in direct light, particles from the light strike a receptor, causing it to turn, verifying that light is indeed a particle with mass.

Lenses
Week 20

Lenses refract or bend light. Convex lenses make objects look bigger; concave lenses make objects look smaller. (Concave looks like a cave.)

What is the difference between reflecting and refracting?
Describe a convex lens.
A magnifying glass is just a lens. Which type? Concave or convex?

Introduction

Two properties of light are that it can **reflect** and **refract**. **Lenses** refract, or bend light. Water, glass, and other objects also refract light. Anything that changes the direction of light is refracting it. **Mirrors** reflect, or bounce, light.

A **lens** is an **optical** device that bends light in a certain way. Basic lenses come in two kinds; they differ in how they bend, or refract, light. A **concave lens**, which has sides that "cave" in, scatters light. The image you see through a concave lens looks smaller. A **convex lens** is thicker at the center. It focuses light on a point called the focal point. Your children may have discovered that if they use a convex lens, otherwise known as a magnifying glass, to focus sunlight, they can possibly light things on fire. Microscopes, cameras, telescopes, and eyeglasses use lenses. Concave lenses are often used with convex lenses to help give convex lenses sharper images. Most eyeglasses have combinations of concave and convex curves.

Mirrors reflect light. They can also be concave or convex, but the kind you are most familiar with are flat. The angle light entering a mirror, called the **angle of incidence**, is the same as the angle reflecting off or leaving the mirror. This is the **angle of reflection.**

The architects of the great cathedrals of Europe understood much about light. The beautiful stained glass not only created beautiful pictures, but it changed the light entering the cathedral. They understood how to mix pigments to create the vivid colors that have not lost their brilliance even today. Often, the edges of the glass were beveled, refracting the light for even more beautiful patterns. The development of the clerestory design increased the amount and quality of light shining into the cathedral.

Supplies

- **Lenses**—an assortment, or one concave and one convex lens per student in class
- **Mirrors**—one per student in class
- **2 Flashlights**—LED works best because it has a more focused beam.
- 2 pieces of white **cardstock** for mirror game targets

Procedure—Part 1

- Pass out lenses.
- Give students a few moments to experiment with the different lenses to gain familiarity.
- Ask students, "What is the difference between the two lenses?" (One makes things bigger; one makes things smaller, or fuzzier.)
- If you can take students outside in the sunlight, allow them to try to focus the sunlight into the brightest possible point on the sidewalk.
- When they find this, they have found the focal point of that lens.

Procedure—Part 2

Review that what we have just experienced is refraction. Light can also reflect, which we will observe next. Keep reviewing terminology until it is mastered.
- Now, pass out the mirrors. Shine flashlight on your mirror while having students observe that the light bounces off at the same angle in which it comes.

- Divide class into two groups. Have students experiment to see how many reflections they can achieve from a beam of light.
- Mirror game: See how many reflections you can achieve from a beam of light.
  - Begin by shining the light on one mirror.
  - Have the holder of that mirror try to reflect it to the next mirror.
  - One student should hold the cardstock target and see if the holder of the last mirror can shine the light on the cardstock target.

- Bring the two groups of students back together to see how many students' mirrors can be included without losing the beam.

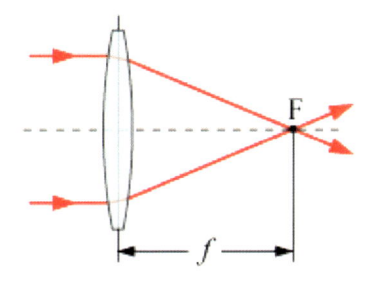

Convex Lens
Light waves *converge* in front of lens.
Notice focal point in front of lens.

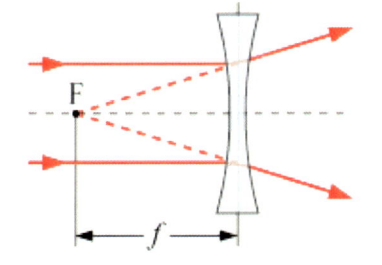

Concave Lens
Light waves *diverge* in front of lens.
Notice focal point behind lens.

Bonus

Set up a mirror maze. Select a target and place black paper on it; then, attempt to set up three or four mirrors so when you turn on the flashlight it will go through the course and end at its destination. Be sure to place obstacles to make it more challenging.

"Sometimes in our lives tears are the lenses we need to see Jesus."

*Pope Francis
Morning Mass
April 2*

Prisms
Week 21

The spectrum from longest to shortest wavelength is: red, orange, yellow, green, blue, indigo, and violet. ROY G BIV

Who is the famous man who helps us remember the order of the rainbow?
What is an example of a longer wavelength than visible light?
What is an example of a shorter wavelength than visible light?

Introduction

The visible part of the **electromagnetic spectrum** is **light**. We are familiar with it in the colors of the rainbow. The different colors are different wavelengths of light. We are familiar with other parts of the electromagnetic spectrum as well; we just don't think about it. **Microwaves** are long wavelengths. They are still part of the electromagnetic spectrum, just longer than red and not visible. **X-rays** are also part of the spectrum, shorter than blue, and not visible.

White light from the sun contains all the colors of the **spectrum**. A **prism** is a transparent solid, usually made of glass with three solid faces. Light passing through it will be **refracted**, or bent. How much a substance bends light is known as its **index of refraction**. The index of refraction in a medium is different for different wavelengths of light. After entry of white light at the first boundary of a triangular prism, there will be a slight separation of the white light into the component colors of the spectrum. After exiting the triangular prism at the second boundary, the separation becomes even greater. Because violet light, with a short wavelength, refracts or bends more than red light, the white light is separated into the colors of the rainbow. This is called **dispersion**.

Violet will always be at the bottom of a rainbow because it has the shortest wavelength. The light aligns itself, wave by wave, causing the longer red wavelengths to take up more length than the blue, resulting in the familiar curve.

Roy G. Biv stands for the first letter of each word: Red, Orange, Yellow, Green, Blue, Indigo, Violet.

Supplies

- **Prisms**—several, one per student in class, if possible.
- **Light source**—direct sun from a window will work, but going outside would be better.
- **White paper**

Procedure

- Ask students if they know which color is always at the bottom of a rainbow. (blue) (violet)
- If they do, great, but do they know why?
- Explain that all waves have a wavelength. A number of waves with a shorter wavelength would be shorter than the same number of waves with a longer wavelength, right?
- "Violet has the shortest wavelength, so . . .where do you predict it will end up?"
- "That's right, at the bottom of the curve, which is the shortest distance."
- Proceed to the area where sunlight is available.
- Pass out prisms. If they are glass, you might wish to stand in a grassy area, as falling to a concrete surface will result in chipped prisms.
- Show students how to hold them to the light to get a good rainbow.
- It takes some experimentation to get it at the correct angle. Be patient, and keep trying. Colors will be more vivid if cast on a sheet of white paper.
- Observe—"Is the same color always on a certain side?"
- "What happens if you turn your prism a different direction?"
- Continue discussing, experimenting, and having fun!

You have learned in art class that the primary colors are red, yellow, and blue. By using these three colors in different Amounts, you can create any other color.

Light is different. On a television screen, the three colors red, green, and blue can make any other color of light.

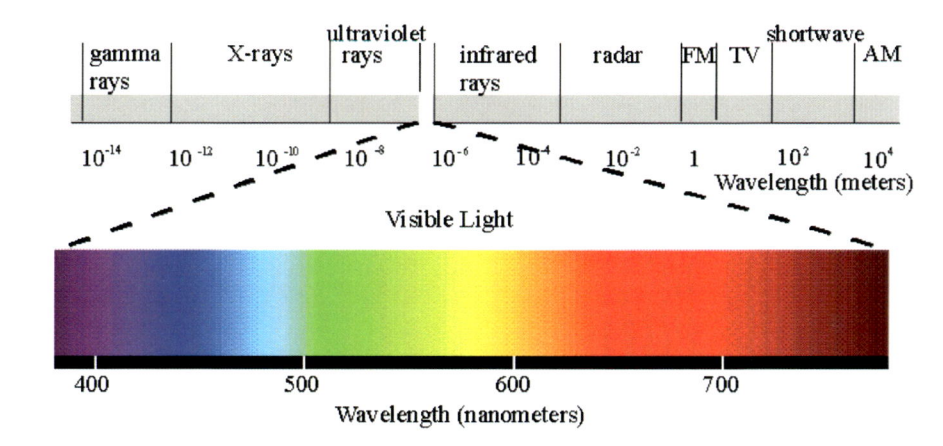

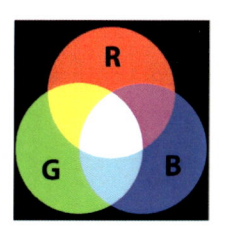

With light, colors are additive. All colors together make white.

Bonus

Students can make their own chart comparing the wavelength of electromagnetic energy across the spectrum with common objects the same size as each wavelength range. Bacteria 300-400 nm, wood smoke 100 nm, . . . 10m for a garage.

Sound & Light - Week 21

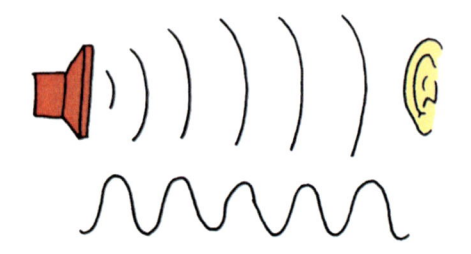

Matter or Energy?
Week 22

Sound is vibrations or waves traveling through matter.
Mach 1 is the speed of sound through air.
Mach 1 equals 770 miles per hour.*

What is the height of a wave called?
What is the frequency of a wave?
What are two types of waves?

Introduction

A **wave** is a phenomenon that travels through a **medium**. A wave is **energy**, not **matter**. The effect of the wave on the surrounding medium is temporary, not permanent.

When a wave moves in an up-and-down or side-to-side motion, it is a **transverse wave**. The height of the wave is its **amplitude**. We learned something about wavelength last week. The **frequency** of a wave is the number of times it happens per second. Longer waves occur less often, so they have a lower frequency. Shorter waves occur more often, and so, they have a higher frequency.

Waves can also travel through a medium like a Slinky. These waves are called **longitudinal** or **compression waves**. This is what happens inside a musical instrument. The air is compressed in waves like the compression motion of a Slinky.

Light and sound are two types of transverse waves, but there are more waves than just the electromagnetic spectrum we learned about last week. Other types of waves occur after an earthquake and in the ocean.

AM and FM radio use frequency and wavelength in their signals. Antennas for different wavelengths are different sizes. The wavelength of an average FM wave is about 1 meter; which is why car radio antennas are approximately 30" long. The military has used extremely low frequencies for some submarine communication. One of these antennas in Michigan was 32 miles long, because it was designed to receive a signal with a wavelength of 32 miles!

Supplies

- **Rope**—a jump rope will work fine.
- **Coil spring**—Slinky is perfect.
- **Wide glass bowl**
- **Plastic wrap**—good quality
- **Rice**—about 1 tablespoon

Teacher Prep

Stretch plastic wrap tightly over top of bowl like a drum. Some flimsy plastic wraps will not be sturdy enough or cling tightly enough. Make sure yours works before going to class.

*Today's memory works states that Mach 1, which we recognize as airplane speed, is 770 miles per hour. This is an approximation, as the speed of sound varies with temperature and pressure.

Procedure—Part 1

- We will demonstrate two types of waves today.
- The first is a longitudinal wave. Have two students each hold an end of the rope. Have them begin to lift and drop so they get vertical waves from the rope.
- Make the waves as tall as possible. This increases the amplitude of the wave. This is what AM radio uses for its signal. AM stands for Amplitude Modulation, or change in amplitude.
- Now make the wave short but fast. The speed of the wave, or more correctly, the distance between its peaks, is the wavelength, or frequency of the wave. Frequency is how often something happens. If you change, or modulate, the frequency of radio waves, you have Frequency Modulation, or FM radio.
- Now bring out the Slinky. Play with the Slinky until you can get a wave to travel lengthwise through it. Teach the students that this is a longitudinal, or compression, wave. Earthquakes put out this kind of wave. The frequency of these waves can also be measured and can help scientists discover through what type of rock earthquake waves are traveling.

Procedure—Part 2

- "Did you know sound can move objects?" Bring out bowl with plastic wrap on top.
- Place several kernels of rice on top of the plastic wrap.
- Use your voice to make a steady sound, changing pitch until the rice vibrates on the plastic wrap. (Higher might work better, but it is really about finding the resonant frequency of your particular bowl.)
- The sound isn't moving the rice, but the plastic wrap, causing it to vibrate at a particular frequency. If you sing louder (greater amplitude), at the right frequency, the rice should bounce higher.
- Similarly, sound waves vibrate your eardrums, allowing you to hear.
- Experiment with changes in loudness and pitch.

Procedure—Part 3

- Students will want to try everything. Use the last 10—15 minutes of class to allow the students to move among stations, trying all activities themselves.
- Divide the number of students by three (the number of activities).
- Indicate when it is time for students to rotate so they get a chance at everything.

Bonus

Kits to build your own FM radio are readily available. Building one can allow students an opportunity to experiment with different length antennas.

Sound & Light - Week 22

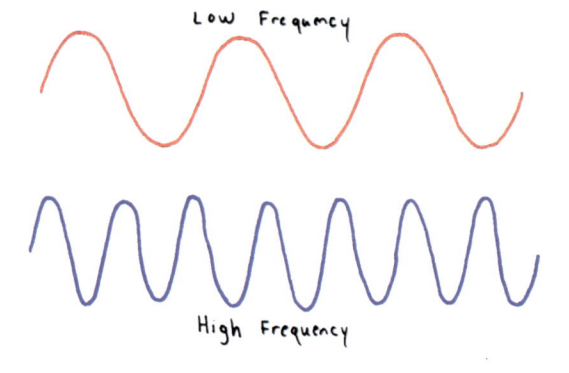

Resonating Rubber Bands
Week 23

Low Frequency

High Frequency

Amplitude is the height of a wave which measures how much. For light, this is how bright. For sound, this is how loud.

Loud sounds have greater _____?
What determines the color of light?
Low sounds have low _____?

Frequency is how many waves happen in a second. This makes light different colors and makes sounds high or low.

Introduction

Low-frequency sound waves have a low **pitch**, like the thickest string on a guitar. **High-frequency** sound waves vibrate at a higher rate, resulting in a higher pitch. The faster the object vibrates the higher its pitch.

Many materials or objects have a specific frequency at which they begin to resonate, or vibrate. For example, air blown through a musical instrument, such as a recorder, resonates at a specific pitch. If you cover holes to change the object, in this case lengthening the tube, it resonates at a lower pitch.

Frequencies unique to an object are called *resonant frequencies*. **Resonant frequencies** can occur on your car, as wind howls around your house, or even on a bridge. Search the Internet for "Tacoma Narrows Bridge" for a large-scale video of the power of resonant frequencies.

Supplies

- **Flat rubber bands**—⅛" and ¼" wide, one of each per student in class
- Sturdy **plastic cups**—one per student in class
- **Straws**—one per student
- **Wineglasses**—8-10 to get a good variety of pitches
- **Water**

Teacher Prep

- Test rubber bands. If they are roundish, they won't work. You need flat rubber bands.
- Become an expert at wineglass music.
- Prefill glasses to a variety of water levels for different pitches.

Procedure—Part 1

- We learned last week that sound waves can cause movement in the plastic wrap and our eardrums.
- This week, we will see how movement can cause sound.
- Vibration is a word for something that moves back and forth quickly.
- Pass out cups, rubber bands, and straws.
- Show students how to stretch rubber bands around the cup lengthwise so the rubber band is across the top.
- Use straw to direct your air across the rubber band.
- Blow, gently at first. It is more important to have the correct alignment than to blow hard.
- If the rubber band moves, students have made it vibrate at its resonant frequency.
- Try the other size rubber band to see if you get a different frequency, or pitch.

Procedure—Part 2

- Now, we will hear some different frequencies.
- Bring out wineglasses. Line them up in order of pitch.
- Wet finger, and rub around rim of glass. A light touch is required. When you get a hum rather than a squeak, you have it!
- Ask students whether the pitch was high, low, or medium.
- Choose another glass. Point out the difference in water level. Before playing, ask students to predict whether the new sound will have a higher or lower pitch than the old sound did.
- Play glass.
- Whose predictions were correct? Why? "Can you come up with a statement that will tell what pitch to expect?" (The more water in the glass, the lower the frequency of the sound waves.)
- You might wish to give students a chance to try their hand at the "wineglass symphony." Remember, fingers need to be wet.

Did you Know?

The pitch on a piano varies with the thickness and length of the strings.

Thick, long strings vibrate more slowly, resulting in a lower pitch.

Thin, short strings vibrate more rapidly, making a higher-pitched sound.

Bonus

Use an electronic tuner to observe and record the frequency of the notes you have learned on the recorder. You can buy these at music stores or as a free app for your smart phone. In what frequency range do the various musical instruments play? Make a chart comparing them.

Squawk, Squawk
Week 24

Light is measured in lumens.
Sound is measured in decibels.
Above 85 decibels can damage your ear.

How many decibels can damage your ear?
How much louder is a 50 dB sound than a 30 dB sound?
What part of our body is a sound receptor?

Introduction

Sound is measured in **decibels (dB).** The decibel scale is not a linear scale. This means that you can't add and subtract decibels like ordinary numbers to get the difference in sound pressure. The chart on the next page shows some common reference sound intensities.

The **eardrum** is a sound pressure receptor. The pressure moves the eardrum. The auditory nerve transfers the message to the brain where it is interpreted as sound.

A **sounding board** is used to **amplify** sound. Our activity today is an example of how a sounding board works. The inaudible vibrations from the string spread through the cup, which amplifies them. Pianos and music boxes use wood as a sounding board in the same way to make the instrument louder.

Acoustics refers to how sound behaves in a space. We say a concert hall has great acoustics when the sounds travel well from the stage to the audience. Acoustics were a very important part of architectural design before the invention of electronic speakers. The round apse of churches, such as the great cathedrals of Europe was designed to focus and amplify the sound coming from the priest and deacons. Especially in the traditionally rear-facing position, the beautiful chants of the Mass were projected toward the congregation.

Supplies

- **Plastic drinking cup**—one per student
- **Yarn or cotton string** (nylon string will not work well)—one 20" per student
- **Nail**—two to three per class
- **Scissors**—one per student
- **Paperclip**—one per student
- **Water** in a small bowl—one per class
- **Paper towels**—one per student

Procedure

- Today, we will make a sound machine with a fun, loud noise that sounds like a chicken.
- Kids love to help. Choose a student helper to pass out each item.
- Begin by cutting a length of string about 20" long for each student.
- Pass out paperclips. Tie one end of the string to the paperclip.
- Pass out cups. Use nail to poke a hole in the bottom of each cup.
- Instruct students to thread the string through the hole so the paperclip ends up on the outside of the cup.
- Pass out paper towels and scissors. Students should cut their paper towel to be about the size of a dollar bill.
- Pass around water, allowing students to slightly dampen their paper towel. Do not immerse them; they will be too wet.
- Fold paper towels in half and wrap around the string near the cup. Squeeze and pull downward to make your chicken sound.
- Short jerks will result in chicken-like squawks!
- Remember, not too loud—over 85 decibels can damage your ears.

20 dB	Whisper or ticking watch
40 dB	Average home or quiet office
45 dB	Amount needed to awaken an average person
60 dB	Conversational speech
70 dB	Average radio
80 dB	Alarm clock, electric shaver, vacuum cleaner
100 dB	Passing truck, lawn mower
120 dB	Fireworks display, damage occurs almost instantly

Did you Know?

The smallest bone in the human body is the stirrup bone, in the ear.

Bonus

Try different types of string in the above project. Does the type of sound change? What about different sized cups?

Scientific Method Lab Report

Name: _____ Date: _____

Question: _____
_____

Procedure/Experiment (What you will do or did you do?):

Materials Used (Draw or list the materials.):

Hypothesis (What you think will happen and why?):

Observation and Analysis (What happened?):

Conclusion (What did you learn?):

SCAVENGER HUNT

Can you find each of these leaf types?

VENATION

Pinnate — secondary veins paired oppositely

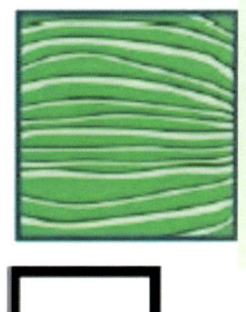

Parallel — veins arranged axially; not intersecting

Palmate — several primary veins diverging from a point

ARRANGEMENT

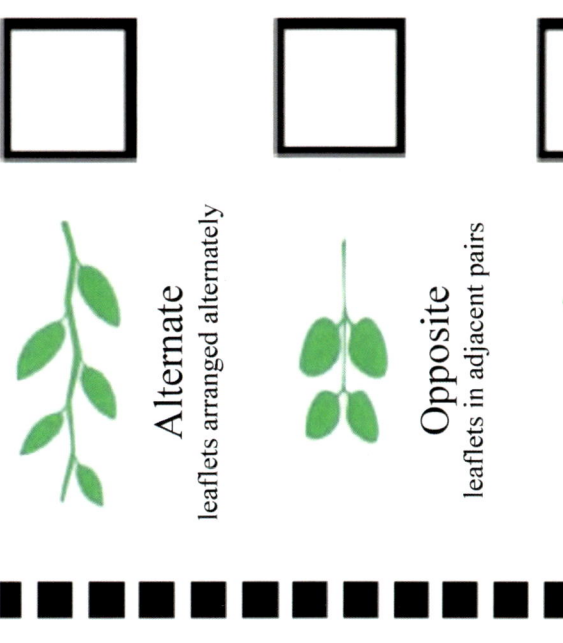

Alternate — leaflets arranged alternately

Opposite — leaflets in adjacent pairs

Whorled — rings of three or more leaflets

COMPOUND

SIMPLE

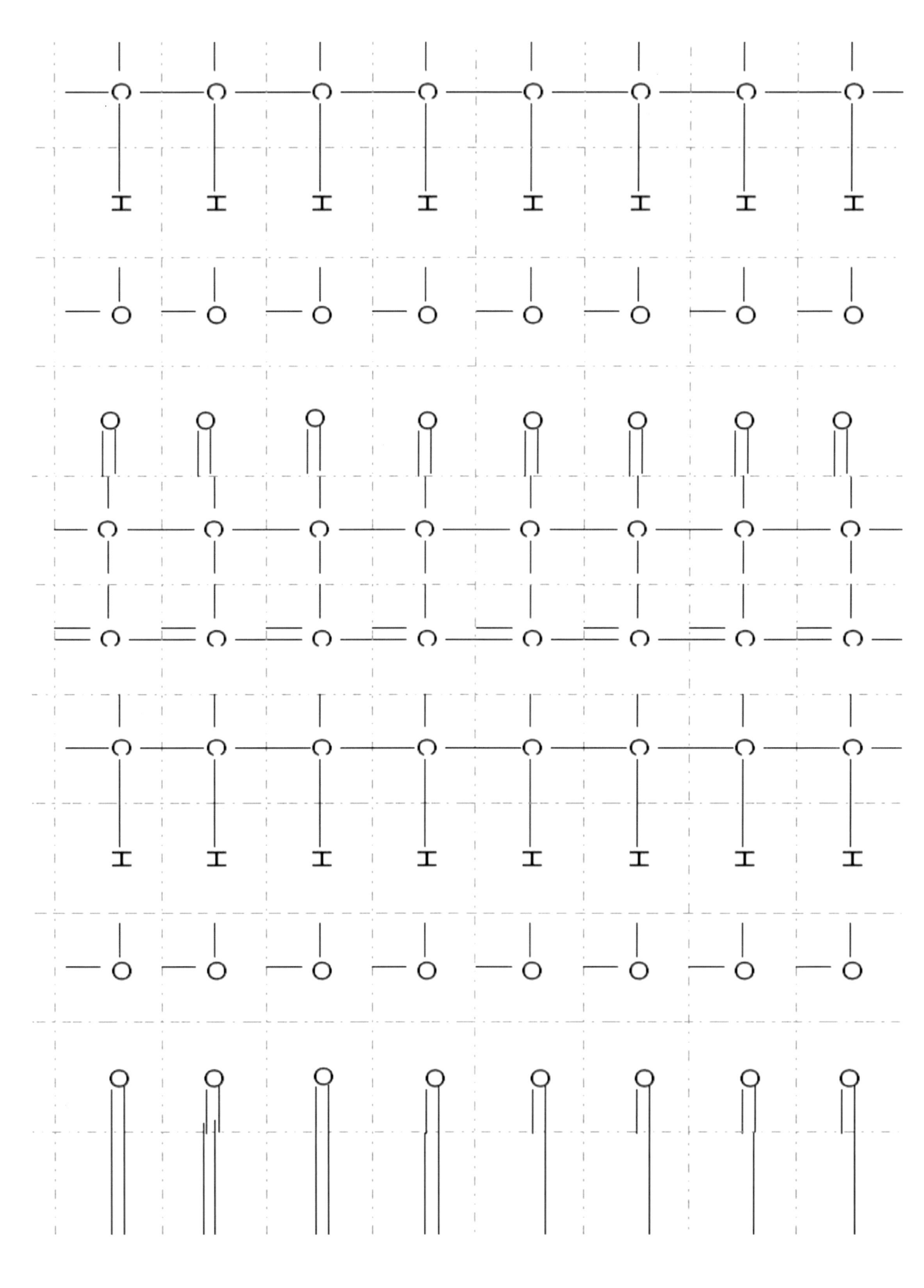

Water	**Glucose**
H_2O	$C_6H_{12}O_6$
Formaldehyde (Disinfectant)	**Hydrogen Peroxide**
CH_2O	H_2O_2
Vinegar (acetic acid)	**Carbon Dioxide**
$C_2H_4O_2$	CO_2

Space Grid for Mapping Constellations

	a	b	c	e	f	g	h	i	j	k	l	m	n	o	p	q	r	s	t	u	v	w	x	y	z
1																									
2																									
3																									
4																									
5																									
6																									
7																									
8																									
9																									
10																									
11																									
12																									
34																									
14																									
15																									
16																									
17																									
18																									
19																									
20																									
21																									
22																									
23																									
24																									
25																									
26																									
27																									
28																									
29																									
30																									
31																									
32																									
33																									
34																									
35																									
36																									
37																									
38																									

Constellation Coordinates

*Print and cut apart,
or
write on whiteboard.*

Cephus

(G, 6) (E, 10) (I, 12)
(J, 8) (O, 11)

Ursa Major

(M, 37) (Q, 34) (R, 34)
(U, 33) (W, 35) (Z, 32)
(X, 30)

Draco

(B, 33) (C, 30) (E, 32)
(D, 34) (B, 24) (C, 22)
(F, 24) (G, 22) (G, 28)
(G, 30) (I, 31) (N, 30)
(R, 27) (U, 27)

Ursa Minor

(R, 17) (O, 18) (N, 20)
(M, 22) (K, 22) (L, 25)
(N, 25)

Cassiopeia

(L, 1) (K, 4) (O, 4)
(S, 5) (R, 2)